Evolutionary ecology of neotropical freshwater fishes

Developments in environmental biology of fishes 3

Series Editor
EUGENE K. BALON

Evolutionary ecology of neotropical freshwater fishes

Proceedings of the 1st international symposium on systematics and evolutionary ecology of neotropical freshwater fishes, held at DeKalb, Illinois, U.S.A., June 14-18, 1982

Edited by
THOMAS M. ZARET

Reprinted from *Environmental biology of fishes* 9 (2), 1983
with addition of three more papers from the conference
and synopses in Spanish and Portuguese

1984 **Dr W. JUNK PUBLISHERS**
a member of the KLUWER ACADEMIC PUBLISHERS GROUP
THE HAGUE / BOSTON / LANCASTER

Distributors

for the United States and Canada:
Kluwer Boston, Inc., 190 Old Derby Street,
Hingham, MA 02043, USA
for all other countries:
Kluwer Academic Publishers Group, Distribution Center,
P.O. Box 322, 3300 AH Dordrecht, The Netherlands

ISBN 978-90-481-8524-5

Library of Congress Catalog Card Number; 83-22238

Library of Congress Cataloging in Publication Data

International Symposium on Systematics and Evolutionary
Ecology of Neotropical Fishes (1st : 1982 : DeKalb,
Ill.)
Evolutionary ecology of neotropical fishes.

(Developments in environmental biology of fishes ; 3)
Synopses in Spanish and Portuguese.
"Reprinted from Environmental biology of fishes 9 (2),
1983 with addition of three more papers from the confer-
ence."
Includes index.
1. Fishes--Latin America--Evolution--Congresses.
2. Fishes--Latin America--Ecology--Congresses.
I. Zaret, Thomas M., 1945-   . II. Title. III. Series.

QL618.2.I58 1982   597'.052632'098   83-22238

Cover design: Max Velthuijs

Preface

In August of 1980, near the whistlestop of Maltby, Washington, Don Stewart and I met in my rented house trailer to sketch a proposal to the National Science Foundation. Our goal was simple: to request from the Foundation air fare and per diem for approximately 20 Latin American scientists to attend a workshop entitled the 'Systematics and Evolutionary Ecology of Neotropical Freshwater Fishes' that would follow the 1982 ASIH (American Society of Ichthyologists and Herpetologists) meeting. We had presented an initial outline for our proposal to a number of colleagues in June of 1980 at the ASIH meeting at Texas Christian University in Fort Worth, Texas. The steering committee for the workshop, consisting of a dozen senior scientists in systematics and ecology of tropical freshwater fishes, met and encouraged us to proceed with submitting a proposal. In addition, the Field Museum in Chicago was making a contribution in order to facilitate this group of scientists. In short, we felt our objectives were correct, we had the backing of the scientific community, and our request was a modest one. We hoped the document would transit the Washington bureaucracy, be mailed to outside reviewers who would give it excellent reviews, and be returned to the NSF Systematics Panel in Washington where the Foundation would give its stamp of approval and send the requested money. We could then invite our Latin American colleagues to the workshop.

It was not until the very end of December, 1981, that we learned our proposal had been turned down. We then hurriedly prepared another proposal to a different government agency in Washington, but discovered that funds were no longer available to support workshops. It was a crisis time of federal budgetary reductions in Washington, resulting in uncertainty for basic research; the new federal budget held no hope for us. Our next approach in early 1982 was to seek less traditional sources. Surely, we would plead, the U.S.A., a wealthy nation, can fund our proposal if only as a gesture of support to foreign scientists. Somehow, however, we seemed to miss deadlines, fall in-between the cracks, and miss the right connections. It was not until May, 1982, several weeks before the proposed workshop, that we realized we could not find any funds for bringing Latin American scientists to the U.S. The programs for the meeting had been printed, the meal coupons, banquet tickets, and all the other amenities that come with a professional meeting were ready, but we had no Latin American ichthyologists as participants. Some abstracts were being received by the program organizers, but without U.S. funding support no one was certain how many Latin Americans would be able to attend. Instead, we might hold a meeting on the Systematics and Evolutionary Ecology of Neotropical Freshwater Fishes without scientists from below the U.S. border. Despair was ours to think that all our efforts went for naught.

Only a little after that date, however, the rumors began. Using other funds, Francisco Mago might be able to attend in spite of recent economic problems in Venezuela; further rumors came that Labbish Chao and some colleagues might arrive from Brazil. But it was all rumors. Not until the very day of arrival at the meetings in DeKalb, Illinois, on the Northern Illinois University campus, did we realize how many Latin American scientists had made the financial commitment to attend. Some were already studying in the U.S., but the majority came on their own from South America, somehow trusting the organizers of the program, and expecting that they would be the one, or perhaps two, Latin Americans in attendance. At final count 15 Latin American scientists attended the conference. In addition, the sections on neotropical fish studies were so popular that extra evening sessions had to be added to accommodate

Thomas M. Zaret (ed.), Evolutionary ecology of neotropical freshwater fishes. ISBN 90 6193 823 6

the total of 45 presentations (about 25% of all papers given).

With each day of meetings, the feeling of familiarity and scientific rapport increased among members from different countries. By the final two days, when the only participants remaining were those attending the special workshop sessions, there had evolved a new spirit in the ASIH, as encapsulated in the last evening. There, in the dormitory lounge of the University, long after all the papers had been presented, there was a non-stop orgy of slide presentations – the kind of performance which only devout ichthyophiles can enjoy. And in the background was continuous music of Latin America – Colombia, Ecuador, Brazil, Venezuela – contributed by various participants. The ebulience was contagious. This was truly the first *international* meeting of ASIH.

Something must be said about the review process for the papers in this Proceedings. All manuscripts were reviewed by two and sometimes three reviewers, the majority of whom were participants at the ASIH meetings. As special editor, I also reviewed every manuscript in its entirety. Following this process, manuscripts were returned to the authors for revisions. In some cases there were three or four revisions before a manuscript was accepted. Several manuscripts were not accepted for this volume. The selection of papers was not intended to cover all bases. The study of evolutionary questions about the diverse tropical fish fauna is in its infancy. Papers were chosen that were scholarly, interesting and sometimes controversial. The goal was to convey to the reader the excitement of studying these natural communities and their inhabitants.

The volume is introduced by the remarkable world of scale eaters, primarily a tropical phenomenon, Ivan Sazima's unique contribution. Sazima explores questions about the polyphyletic behavior of scale eating fishes, and presents an original hypothesis about the origin of scale eating – from social interactions. The next three papers are related in their focus on feeding methods of fishes that predominate in these communities.In Mary Power's paper, among her findings is that choice of grazing response by loricariid fishes is determined to a great extent by their investment in predator avoidance strategies. Their response to seasonal variation of food availability is also highlighted. The contribution by Paul Angermeier & James Karr focuses on the relationship of fish communities to the stream environment. The extensive data base allows them the unique opportunity to compare tropical streams with temperate streams, on which they also are expert. And let's not forget the detritivores. The relative importance of detritivory, as discussed by Stephen Bowen, varies from system to system, but estimates give 50% or more of community ichthyomass to these fishes. Yet they are often overlooked in ecological studies.

The next six papers cover a broader spectrum. Don Kramer asks the question: Why is there such diversity in respiratory modes in fishes, essentially putting the shoe on the other foot of those who have implied that there may be *one* most efficient way of handling this problem. The contribution by Bruce Turner et al. raises questions about the relationship between polymorphism and phenotypic plasticity in fishes. Their studies on the Goodeidae imply that what may pass for species flocks could really be species with great morphological diversity, even to dentition. The chapter by John Endler presents a great wealth of information available on poeciliids, with emphasis on natural sexual selection of guppy color patterns. The following three papers appear only in the sibling hardcover volume 3 of Developments in Environmental Biology of Fishes, together with abstracts in Spanish and Portuguese of all the 10 papers published. In Hugo Campos' paper on biogeography we have a strong voice interpreting recent discoveries in terms of an alternative view of biogeography. His interpretations of the origin of the Southern Hemisphere salmonids will spur a great deal of healthy controversy. Thomas Zaret and Eric Smith continue thoughts on the role of time with two questions which have been of concern in ecology, namely, diet overlap and its relationship to morphological diversity. The final paper is by the Grande Dame of neotropical fish ecology, Rosemary Lowe-McConnell, whose contributions on freshwater tropical fish communities are known throughout the world. In this paper, however, she moves in a different direction, concerning herself with questions basic to fisheries management. A controversial conclusion in this contribution is that the continuous decline in fish stocks in South American fisheries may be due more to man's recent pressures of deforestation and related activities, than overexploitation by fishing and fishery techniques.

Many people contributed to what has now become the Symposium issue of Environmental Biology of Fishes and a sibling hardcover volume. These include the Steering Committee members: Jonathan Baskin, Jim Böhlke (now deceased), John Endler, William Fink, John Lundberg, Francisco Mago-Leccia, Náercio Menezes, Robert R. Miller, Don Stewart, Cam Swift, Stanley Weitzman, and myself. These individuals gave advice and support, and most attended the 1980 (ASIH) discussion at T.C.U. Other individuals who contributed to this Symposium Volume include: Eugene Balon, Dave Greenfield, Jeanette Pederson, and Barbara C. Petersen. Don Stewart also selected samples of illustrations from Steindachner (1879, 1880, 1882) to fill some of the blank pages. I thank Mark Andersen for hours of proofreading manuscripts, and catching my mistakes and omissions, Eric Volk for laboring valiantly on the maps and innumerable small services, Otto Infante, Venezuela, for the Spanish translations, and Ivan Sazima for the Brazilian translations of the abstracts. The logo is a fish (a goby) candle holder by an unknown neotropical Indian from the collection of primitive art owned by Eugene Balon.

Thomas M. Zaret

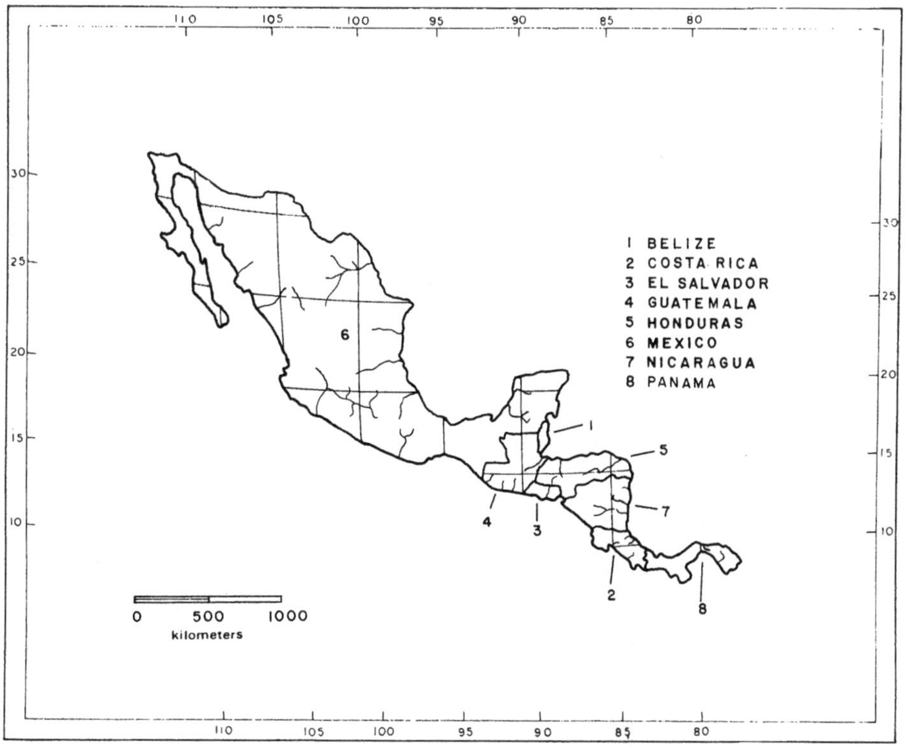

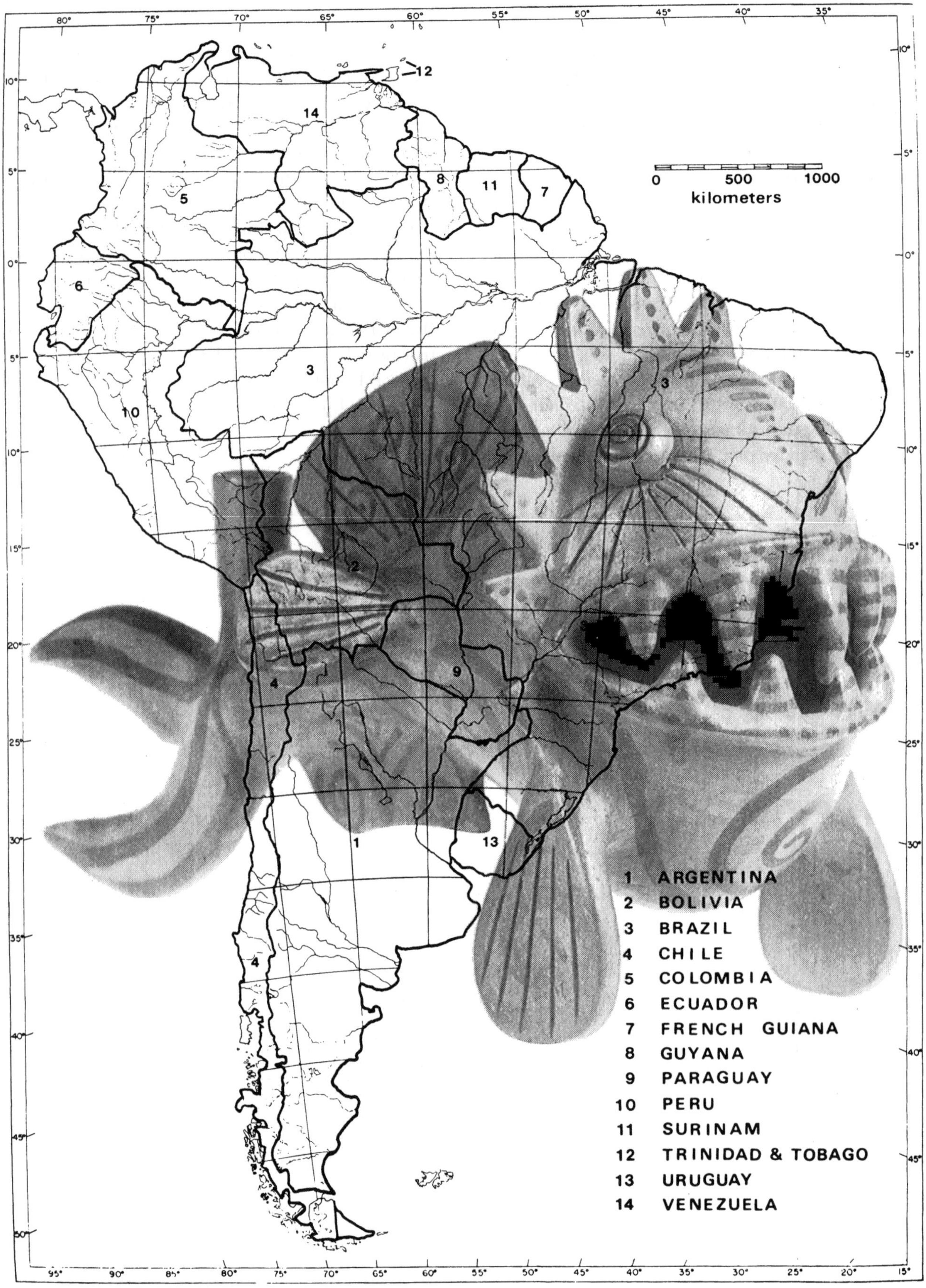
0 500 1000
kilometers
1 ARGENTINA
2 BOLIVIA
3 BRAZIL
4 CHILE
5 COLOMBIA
6 ECUADOR
7 FRENCH GUIANA
8 GUYANA
9 PARAGUAY
10 PERU
11 SURINAM
12 TRINIDAD & TOBAGO
13 URUGUAY
14 VENEZUELA

Scale-eating in characoids and other fishes

Ivan Sazima
Departamento de Zoologia, Universidade Estadual de Campinas, 13100 Campinas, Sao Paulo, Brasil

Keywords: Feeding habits, Lepidophagy, Evolution, Behavior, Predation

Synopsis

Scale-eating is known for several unrelated fish groups, but few data are available on the habits of most species. General habits and feeding behavior of some lepidophagous characoids are presented and compared to other scale-eating species. The diversity of morphology, habits, and behavior of scale-eating fishes is great, and few patterns are shared by the specialized scale-eaters. Except for modified teeth, no morphological characteristic permits identifying a fish as a specialized lepidophage. Hunting tactics consist mainly of ambush, stalking, or disguise (aggressive mimicry). Scale-removal may be accomplished by a jarring strike with the snout, generally directed at the prey's flank, or by biting or rasping. The mode of scale-removal seems to reflect primarily the disposition of the jaws and the teeth. Scales are swallowed directly if taken in the mouth; if not, they are gathered as they sink, or picked up from the bottom. Scale-eating is probably a size-limited habit. Specialized scale-eaters rarely exceed 200 mm, most ranging near 120 mm. Some species eat scales only when young; most take other food items in addition to scales. Scale-eating habits probably arose from trophic or social behaviors. These are not mutually exclusive and, indeed, may have acted together during the evolution of lepidophagy. Suggested trophic origins include scraping epilithic algae, modified piscivory, and necrophagy. Social origins include intra- and interspecific aggressive behavior during feeding.

Introduction

Various fish species have adopted the habit of eating particular parts of other living fishes, including scales, skin, fins, gill filaments, blood, chunks of the body and even eyes (see Fryer et al. 1955, Roberts 1972, Curio 1976). Among these modes of mutilation, scale-eating is perhaps the most widespread, found in a number of fish species of several unrelated taxa from diverse habitats, in both freshwater and the sea.

Among the freshwater fishes, lepidophagous species have been found in five African genera of Cichlidae (Fryer & Iles 1972, Eccles & Lewis 1976, Liem & Stewart 1976), seven genera of neotropical Characidae (Roberts 1970, Géry 1980, Goulding 1980), an Asian genus of Schilbeidae (Whitfield & Blaber 1978), one genus of Ariidae in Australasia (Roberts 1978), and at least three genera of neotropical Trichomycteridae (Baskin et al. 1980). The supposed scale-eating habits of *Oedemognathus* (Roberts 1970), gymnotiform Apteronotidae, have yet to be confirmed. Among marine fishes, scale-eating is known mostly for juveniles. Lepidophagous species have been found in one genus of Ariidae in the W. Atlantic (Hoese 1966), one genus of Blenniidae in the tropical Pacific (Losey 1972a, 1978), two genera of Carangidae in the W. Atlantic

Thomas M. Zaret (ed.), Evolutionary ecology of neotropical freshwater fishes. ISBN 90 6193 823 6

and Pacific (Major 1973, Sazima & Uieda 1980), two genera of Triacanthodidae in the Pacific (Mok 1978), one genus of Labridae in the tropical Pacific (Losey 1972b), one genus of Teraponidae in the Pacific (Whitfield & Blaber 1978) and two genera of Kyphosidae in the Pacific (DeMartini & Coyer 1981). Thus lepidophagous habits are presently known in at least five freshwater and seven marine fish families, mainly from tropical regions.

Lepidophagy is regarded as a derived, highly specialized habit, and scale-eating fishes are often morphologically and behaviorally specialized for feeding (e.g. Roberts 1970, Fryer & Iles 1972, Major 1973, Whitfield & Blaber 1978, Sazima 1980). In some instances it is associated with aggressive mimicry (Trewavas 1947, Sazima 1977), suggestive of its evolutionary complexity.

In this paper I present data on scale-eating in some neotropical characoid species and compare them mainly with African cichlids and some marine species. Most of the study sites and general procedures are described elsewhere (Sazima 1977, 1980, Sazima & Uieda 1980, Sazima & Machado 1982).

I will emphasize the relationships between morphology and behavior, and indicate patterns apparently common to most lepidophagous species. Also, I will indicate further study which may lead to a better knowledge of scale-eating and its evolution.

Scale-eating characoid genera

Characoids constitute the largest group of South American fishes, the number of species approaching one thousand (Géry 1977). The group has undergone extensive adaptive radiation resulting in an unparallelled diversity of diets and morphologies. The following genera are known to be specialized scale-eaters: *Roeboides*, *Exodon*, *Roeboexodon*, *Probolodus*, and *Catoprion* (Roberts 1970) and the recently described *Bryconexodon* (Géry 1980). Four representative scale-eating characoids are shown in Figures 1–4. A few *Serrasalmus* species (piranhas) practice lepidophagy to a certain extent, but rely mainly on fin-eating (Roberts 1970, Goulding 1980). *Roeboides* is the only polytypic genus of specialized scale-eaters, being widespread throughout the Neotropics and present in almost all major river basins. The other lepidophagous genera are monotypic and have a more restricted range.

All specialized scale-eating characoid genera belong to the large family Characidae (including Serrasalminae). The classification and relationships of characids are insufficiently known to provide information on phyletic relationships of lepidophages. Most lepidophagous genera are placed in distinct tribes (e.g. Géry 1964, 1972, 1977, 1980). In contrast, genera such as *Roeboides* and *Exodon* have been placed together in the same subfamily, Characinae (e.g. Géry 1959, 1977, Roberts 1970), notwithstanding their different appearance (Fig. 1, 3). In fact, preliminary morphological and osteological studies by Naércio A. Menezes (personal communication) suggest that *Exodon* has more affinity with the Tetragonopterinae than with the Characinae, and this seems also to be supported by several aspects of its behavior (Sazima 1980). *Roeboexodon* is sympatric with *Exodon* in parts of its range, and in the Rio Araguaia, Goiás State, Brazil, they seem to be syntopic. The recently described *Bryconexodon* (Géry 1980) probably belongs to the same phyletic assemblage. *Probolodus* was described in the Cheirodontinae but some authors prefer to place it among the Tetragonopterinae (Roberts 1970, Géry 1980). It is clear that one of the most important steps in the study of the scale-eating habits in characoids rests in clarifying the systematic relationships among the species involved (Sazima 1980). The value of such a procedure is illustrated by the work of Liem & Stewart (1976) on the African cichlid *Perissodus*, and the predictions of Baskin et al. (1980) on scale-eating habits of trichomycterid catfishes.

Characteristics of scale-eating fishes

Morphology

Specialized teeth seem to be the only external feature characteristic of all scale-eating characoids. At least some teeth are directed forwards and point out of the mouth or are entirely external (Fig. 5–7). These teeth are stout with hypertrophied bases and may be conical and mammiliform, or cuspidate. The outward pointing of the teeth appears unrelated to jaw length, total tooth number, cusp number or the presumed systematic relationships among species. Such a uniform adaptive response is almost certainly due to the relative simplicity of the mouth architecture as compared, for example, with the complex feeding apparatus of cichlids (Liem & Osse 1975). Certain teeth become modified early in juvenile life, e.g. the fourth dentary tooth of *Probolodus* (Fig. 7), which is clearly differentiated in 19 mm young (adults attain 110 mm). This tooth plays a key mechanical role during scale-removal as deduced from tooth wear and replacement (Roberts 1970) and from scale-removing behavior (Sazima 1980).

The position and wear of the teeth, as well as the relative length of the jaws, allow some predictions about the mode of attack and scale-removal. Thus, *Roeboides prognathus* and *R. paranensis* have a projecting upper jaw and many external teeth (Fig. 5). The foremost premaxillary and dentary teeth which point almost straight ahead are invariably worn and sometimes missing. These species seem to strike at prey mainly with their mouth closed (Sazima & Machado 1982). On the other hand, *Catoprion*, with protruding lower jaw has fewer, less projecting teeth and attacks prey with its mouth wide open. *Exodon*, with upper and lower jaws of almost equal length, has some external teeth pointed forwards and strikes at the prey either with its mouth open or closed. Although I have not observed living specimens of *Roeboexodon* and *Brycon-exodon*, I predict from their jaw shape and dentition that the former can remove scales with its mouth closed whereas the latter probably has to open its mouth. In addition, the differential wear of teeth may indicate the right- or left-handedness of the individual, i.e. the tendency to remove scales from the prey using the right or left side of the mouth. I have observed this in *Roeboides prognathus* (D.L. Kramer has informed me of similar findings for *R. guatemalensis*).

Among the African scale-eating cichlids, there are numerous variations involving the form and disposition of teeth (Fryer & Iles 1972). The versatile cichlid mouth has permitted some remarkable adaptations, such as jaws asymmetrically suspended to the right or to the left for striking the prey more effectively found in *Perissodus eccentricus* (Liem & Stewart 1976). In the other scale-eating fish groups the solutions vary from the presence of distinctive outer dentary teeth, tightly spaced, strongly hooked outwards and with spatulate tips, found only in juveniles of some species of the carangid *Oligoplites* and *Scomberoides* (Smith-Vaniz & Staiger 1973, Sazima & Uieda 1980, and Fig. 8) to the long rows of numerous needlelike teeth in the stegophiline catfishes (Baskin et al. 1980).

The size, shape and adherence of scales in scale-eating fishes is often discussed in relation to their feeding habits (Breder 1927, Roberts 1970, Whitfield & Blaber 1978, Sazima & Uieda 1980). Reduction in scale-size seems to be common in most scale-eaters; the assumption is that small and adherent scales reduce intraspecific scale-eating. We have found that *Roeboides prognathus*, *R. paranensis*, *Probolodus heterostomus*, *Exodon paradoxus* and *Catoprion mento* are all capable of removing scales from conspecifics, apparently with the same ease with which they obtain scales from usual prey species (Sazima 1980, Sazima & Machado 1982, personal observations). In *P. heterostomus* I found conspecific scales in 4 out of 69 stomachs examined. My observations on characoids suggest that scale characteristics of these fishes do not actually prevent intraspecific scale-eating, although they may help to reduce damage when attacks actually occur. Behavioral avoidance is probably a major factor which reduces scale-loss or damage during intraspecific encounters (Sazima 1980). On the other hand, in the carangid *Oligoplites saurus* the deeply embedded, needlelike scales and leathery skin actually does prevent damage to skin in cases of intraspecific (misdirected ?) attacks during the pursuit of schooling prey (Sazima 1980).

An additional possible explanation for small, adherent scales is the role they may play in preventing damage from retaliatory actions of the prey. Several predators exhibit counter-adaptations to the defensive actions of their prey (Curio 1976), and it must be noted that scale-eaters may attack prey larger than themselves (Breder 1927, Sazima 1980), against which they have few defenses other than

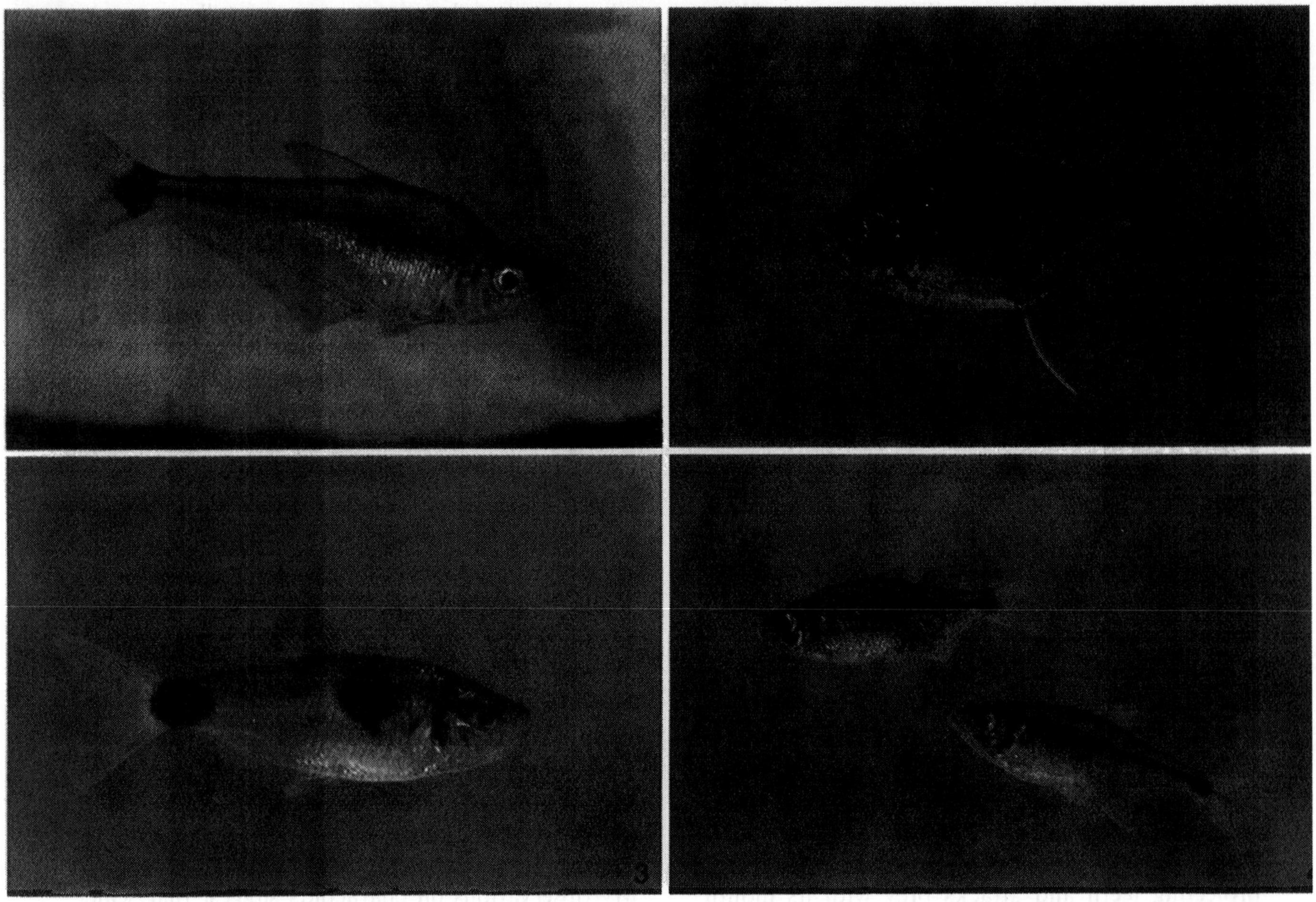

Fig. 1–4. Four representative species of scale-eating characoid fishes: (1) *Roeboides prognathus*, 72 mm standard-length; (2) *Catoprion mento*, 61 mm; (3) *Exodon paradoxus*, 59 mm (note 'twin-spot' pattern); (4) *Probolodus heterostomus*, about 30 mm, shown with its usual prey *Astyanax fasciatus* (upper).

their agility. In a few instances I have observed *Probolodus* being chased by a previously attacked *Astyanax*. Although it usually evaded the pursuer, sometimes the *Probolodus* was overcome and bitten, without, however, resulting in damage to its skin (Sazima 1980).

Among the characoid scale-eaters, the body shape does not appear to reflect any particular adaptation to scale-eating. For instance, the deep body of *Catoprion* (Fig. 2) is the rule among the serrasalmine characoids, as is the somewhat elongated body and anal fin of *Roeboides* among Characinae. On the other hand, genera such as *Exodon*, *Brycon-exodon* and *Probolodus* have a body shape like those found in most tetragonopterine characins (Fig. 3, 4, 11).

The color pattern of some scale-eating fishes has occasionally been discussed in relation to their feeding habits (Roberts 1970, Whitfield & Blaber 1978). The translucent body of some species of *Roeboides* may favor undetected stalking of their prey (Sazima & Machado 1982). The similarity of

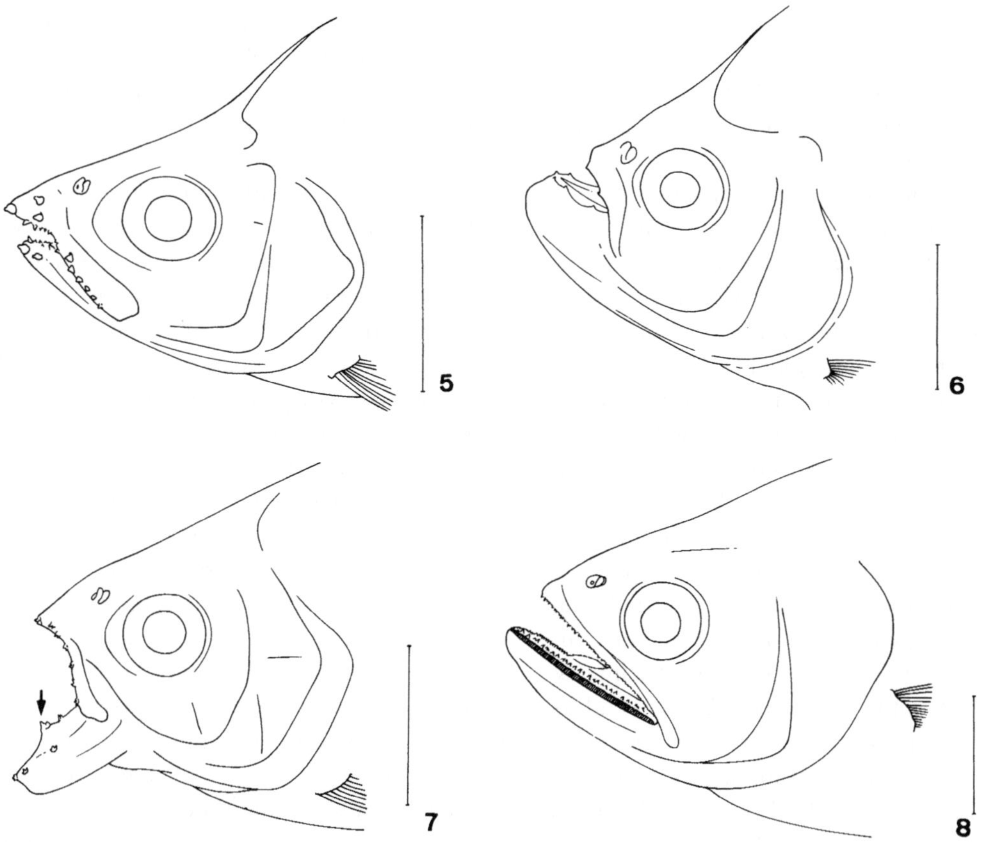

Fig. 5–8. Heads of three characoid and one carangid species of scale-eating fishes to show jaw proportion and teeth types and arrangement (vertical scales are 10 mm): (5) *Roeboides prognathus*, with projecting upper jaw and external, forwards directed, mamilliform teeth (after Sazima & Machado 1982); (6) *Catoprion mento*, with prognathous jaw and few, outwards pointing tricuspid teeth; (7) *Probolodus heterostomus* showing the few, forward directed tricuspid teeth – note prominent position of the fourth dentary tooth (arrow); (8) *Oligoplites palometa*, a carangid, with outer dentary teeth tightly spaced and hooked outwards.

Probolodus heterostomus to *Astyanax fasciatus* (Fig. 4) favors its schooling with this habitual prey species (Sazima 1977, 1980). The 'twin-spot' pattern of *Exodon paradoxus* (Fig. 3) is an exception among scale-eating fishes and has no counterpart among Characidae in general. This is an aggressive and robust schooling species and its dentition is strong and sharp. (It is the only specialized scale-eater I have observed which bites resolutely when handled.) Fishes with such characteristics sometimes possess clearly recognizable intraspecific signal-marks, e.g. reddish belly associated with deep body in *Serrasalmus nattereri* (Markl 1972). The color pattern of *E. paradoxus* may be primarily related to social, intraspecific interactions (Sazima 1980), as already recognized by Lowe-McConnell (1964). Facts suggesting a relation between these body marks and scale-eating will be discussed later.

In summary, scale-eating is characterized by a wide variety of morphological patterns. Aside from dentition, there is no single external feature which characterizes a 'scale-eating type' among the specialized lepidophagous characoids (and perhaps other scale-eating taxa too). The absence of morphological characterization becomes increasingly important when one realizes that to feed on scales, a fish need not necessarily have specialized dentition. Some species of the generalized and omnivorous characoid genus *Astyanax* have been found to ingest scales (Nomura 1975). To recognize the scale-eaters, one has to study feeding habits and behavior.

Behavior

There are few descriptions of scale-eating behavior. About 10 out of the 50 or more known lepidophagous species have been observed actually removing scales of the prey, and these observations have mostly been made on fishes in aquaria. The only studies containing field observations on scale-eating characoids dealt mainly with the general aspects of fish communities (Zaret & Rand 1971, Kramer 1978 for *Roeboides guatemalensis*). I have found lepidophagous characoids in sites ranging from muddy inlets in sizeable rivers (Fig. 9) to very clear pools in small, sluggish creeks (Fig. 10). In scale-eaters, vision seems to be the main sense employed to detect and orient approach to the prey, as well as to attack it and evade possible retaliation. With few exceptions, characids are diurnal and have well developed eyes. *R. prognathus* increases its predatory activity at dusk (Sazima & Machado 1982) and *R. guatemalensis* exhibits crepuscular or even nocturnal predatory behavior (D.L. Kramer, personal communication). Species of this genus have a network of pores on the head, called pit lines (Géry 1966), which I suspect are sensitive structures which allow perception of prey at a distance, a function comparable to that suggested for the branch of the laterosensory canal in the premaxillary of the characoid *Acestrorhynchus* (Menezes 1969). A *Roeboides prognathus* placed in a 50 l tank with one or two prey fishes, removed scales from them in muddy water with visibility not greater than 20 mm (maximum depth at which a white 50 mm long object remained visible), while it did not molest a conspecific, as indicated by stomach contents analysis (Sazima & Machado 1982).

Scale-eating fishes often approach their prey using concealment or mimicry. *Catoprion mento* may stalk or ambush its prey using clumps of waterweeds as cover (Fig. 10). The translucency of *Roeboides prognathus* and *R. paranensis* may conceal their approach (Sazima & Machado 1982). *Roeboides* and *Catoprion* may also linger close to the prey, behaving as if 'not interested' in the prey (sensu Curio 1976) and thereby attack from close quarters. I have also observed this subtle behavior in young *Serrasalmus spilopleura* and *S. marginatus*, two 'mutilating' characoids. The use of harmless fish species as a 'cover' to approach a prey may be one of the predatory tactics of *Probolodus* schooling with *Astyanax fasciatus*. (A similar hunting tactic is used by the skin- and mucus-eating blenniid *Plagiotremus azalea*, which swims 'concealed' within schools of the labrid *Thalassoma lucasanum*, Hobson 1968.) In addition, *Probolodus* and species of the cichlid genera *Corematodus* and *Perissodus* are

Fig. 9. Inlet in the Rio Cuiabá, near Cuiabá, Mato Grosso State, Central Brazil. The often muddy waters of this river harbor three species of *Roeboides*, which may occur syntopically.

Fig. 10. Fishes in a clear pool of a small creek in the Pantanal region near Poconé, Mato Grosso State, Central Brazil. A specimen of *Catoprion mento* appears near the centre of the picture; one *Acestrorhynchus altus* is on the left, below and two *Astyanax bimaculatus* are visible on the right lower corner. Note abundant aquatic vegetation.

aggressive mimics of their prey fish (Trewavas 1947, Sazima 1977, Brichard 1978). By having the prey's appearance (e.g. Fig. 11), they can maneuver to a convenient attacking position. The cichlid *Perissodus microlepis* poises in midwater near rocky slopes, waiting to strike at the unwary victims (Brichard 1978). The carangids *Scomberoides* and *Oligoplites* are open water fishes which often school together with their intended prey (Major 1973, Sazima & Uieda 1980). Young *O. saurus* are greenish and silvery and may adopt the color tone and posture of their intended victims (Sazima & Uieda 1980). On the other hand, young *O. palometa* have a yellowish phase with two dark bars and seem to stalk and ambush prey among marsh grass and other aquatic vegetation. At least one exception to these subtle or deceptive modes of approach is known: *Exodon paradoxus* rushes towards the prey, seemingly without subterfuges (Roberts 1970, Sazima 1980). This uncommon tactic, together with its schooling habits and unique color pattern suggests that *Exodon* may be a group predator similar to the conspicuously patterned teraponid *Terapon jarbua* which also schools when attacking the prey (Whitfield 1979). This teraponid attacks intermittently from behind, thus maintaining the element of surprise (Whitfield & Blaber 1978).

The actual attack varies little among the lepidophagous characoids. Usually, the predator strikes at the flank of a prey from a perpendicular or posterior oblique position (Fig. 12–14). Strikes of *Roeboides prognathus* and *Exodon paradoxus* are mostly caudad (Fig. 13); such a thrust removes scales more easily than one directed against the free edge of scales (Sazima & Machado 1982). *Catoprion mento* strikes most frequently at a right angle to the prey's flank (Fig. 12), maximizing scale-removal by the wide open mouth. A single well directed strike by *Catoprion* removes several rows of scales, leaving a neatly denuded spot on the side of the prey. Field observations indicate that *Catoprion* is actively avoided by its prey. The scale-removing strike of the specialized lepidophagous characoids is jarring and the prey is frequently displaced by the force of impact (Fig. 16). Moreover, most favor attacks on moving fish. The escape reflex of the prey adds to the impact of the strike and this, accompanied in some predators by a lateral head jerk as it bites, helps furnish the force needed to dislodge the scales. After the first attack, the scale-eaters do not follow the prey for a great distance.

Other relationships are important between the attack behavior and the morphology or general habits of a species. When several *Exodon* attack a group of prey fish, their 'twin spot' pattern darkens to its maximum, and they stand out unmistakably

Fig. 11. Interactions in a mixed group of *Probolodus heterostomus* and *Astyanax fasciatus*. The uppermost fishes are *Probolodus* aiming a charge at an *Astyanax*; the lowermost ones are *Astyanax*, one chasing the other.

Fig. 12. Scale-removing attack of *Catoprion mento* towards *Astyanax* prey. The wide mandible can remove several rows of scales in a single strike.

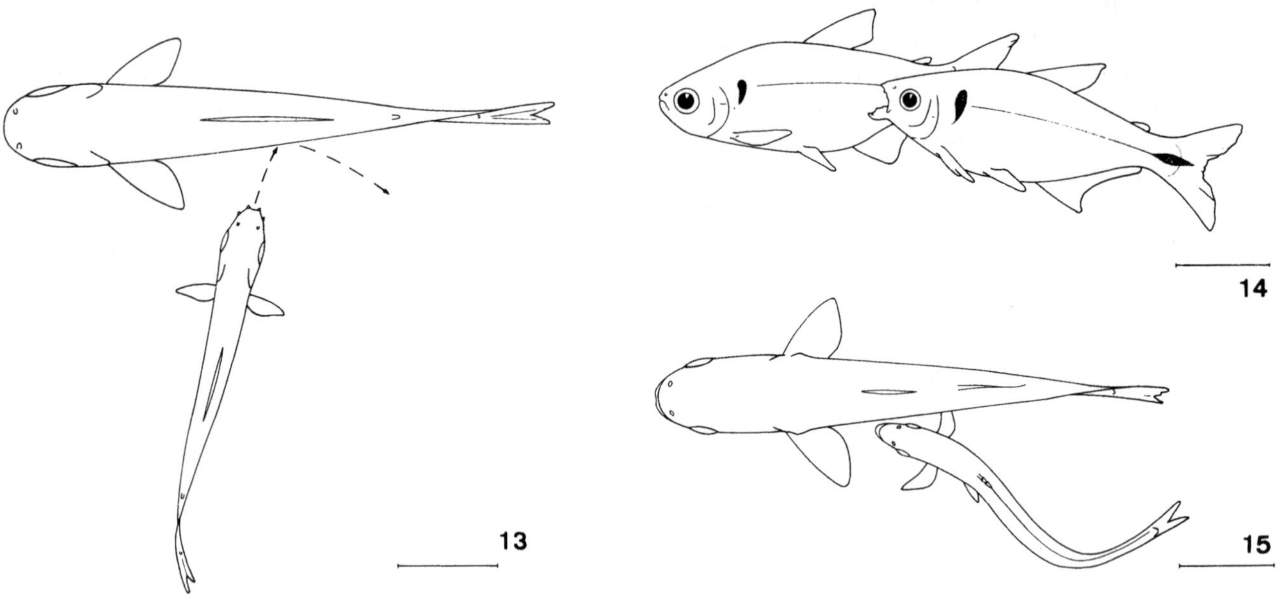

Fig. 13–15. Scale-removing attacks of two characoid and one carangid species (horizontal scales are 20 mm); (13) *Roeboides prognathus* attacking a *Tetragonopterus argenteus* – the predator commonly strikes at the prey with its mouth closed and directs the thrust caudally (after Sazima & Machado 1982); (14) *Pròbolodus heterostomus* biting the flank of *Astyanax fasciatus* – this fish follows the prey and strikes mainly from behind; (15) young *Oligoplites saurus*, a carangid, positioned to strike at the flank of a *Mugil curema* – the wave-like posture is characteristic, and scales are removed with the side of the mandible (redrawn from Sazima & Uieda 1980).

among the milling prey. This possibly allows *Exodon* to easily discriminate the attacked fish and may reduce intraspecific attacks. The oblique, latero-posterior charge of *Probolodus* (Fig. 14) is consistent with its schooling habits and mimetic pattern: *Astyanax* cannot easily discriminate a fish so positioned, and field observations suggest that they mistake the predator for a following companion (Sazima 1980). The cichlid *Corematodus* probably also uses its resemblance to the prey fish to position itself before closing its jaws over the tail of the victim and rasping off the scales as the prey flees (Trewavas 1947, Fryer & Iles 1972). The carangids *Scomberoides* and *Oligoplites* curve the body into an 'S' posture (Fig. 15) and, with the mouth open, strike at the prey's body with the side of the mandible (Major 1973, Sazima & Uieda 1980). This posture is believed to be a requisite both for positioning the mouth so that the teeth can be inserted under the scales (Major 1973) and for providing the force necessary to dislodge the scales (Sazima & Uieda 1980). On two occasions I have observed an individual of *O. saurus* touching the belly of the intended victim with its pectoral fin before the actual scale-removing strike. Physical contact with the host preceding actual feeding seems to be important for cleaning and mucus- and scale-eating activity by the labrid *Labroides dimidiatus* (Losey 1978, 1979).

The dislodged scales are swallowed immediately if taken in the mouth, or gathered as they sink, a behavior which appears in all scale-eaters studied to date. After a strike, *Roeboides*, *Exodon* and *Probolodus* often turn about to pick up the falling, dislodged scales. Scales on the substrate are visually sought and ingested after lifting them with the snout (Fig. 17) sometimes hours after the initial attack.

The fishes attacked probably escape without much harm and recover. (Under aquarium conditions missing scale-rows of *Astyanax* spp. and *Carassius auratus* regenerate in about three to four weeks.) Thus scale removal is less serious to prey fish than fin-clipping, practiced by many *Serra-*

Fig. 16–17. Catoprion mento completing a scale-removing strike on *Astyanax*, showing the prey displaced by the force of the impact (16); *Catoprion mento* seeking for fallen scales, lifting them from the substrate with its wide mandible (17).

salmus species, for a crippled fish is more vulnerable to other predators. Although scale- or fin-eating can be regarded as a parasitic mode of feeding, I consider it predation because of the tactics used.

Prey choice

Most scale-eating fishes attack prey as large as, or larger than, themselves (Breder 1927, Hoese 1966, Major 1973, Whitfield & Blaber 1978, Sazima 1980, Sazima & Uieda 1980). The limits probably are determined by the size of the removed scales (Whitfield & Blaber 1978), as the mouth and esophagus of the predator must be able to engulf them, and by the ability of the scale-eater to approach its victim and to retreat, if necessary (Marlier & Leleup 1954, Whitfield & Blaber 1978, Sazima 1980). Scale-eaters remove scales from diverse fish species (Major 1973, Mok 1978, Whitfield & Blaber 1978, Vieira & Géry 1979, Sazima 1980). Among characoids *Probolodus* preys mostly on *Astyanax* and, to a lesser extent, on *Curimatus* (Sazima 1977). Cichlid species are rarely attacked (Sazima 1977, 1980). *Roeboides prognathus* takes scales mostly from tetragonopterines, *Astyanax* and *Tetragonopterus*, but cynopotamines may also be attacked (Sazima & Machado 1982). In the field, I observed *Catoprion* attacking only *Astyanax bimaculatus*, a ubiquitous fish in the habitat of the former at Poconé, Mato Grosso State (Fig. 10). Vieira & Géry (1979) listed anostomids, hemiodids, and possibly curimatids as prey of *Catoprion* in the Rio Curuá-Una, Pará State. Under aquarium conditions, characoids accept almost every scale-bearing species presented to them, including goldfish *Carassius auratus* (Sazima 1980). Predator preferences (Hoese 1966, Fryer & Iles 1972, Sazima 1977, Sazima & Machado 1982) probably reflect prey attributes such as abundance, spatial and temporal distribution, specific behavior, and the ease of scale removal.

No characoid scale-eater was ever seen removing scales from dead fish, or even freshly killed ones, in aquaria experiments (Sazima 1980). On the other hand, the carangid *Oligoplites saurus* habitually eats scales from dead fish under aquarium conditions. This suggests that *Oligoplites* is to a certain extent a scavenger and that scavenging may have been one of the ways which led to lepidophagy in this genus (Sazima & Uieda 1980).

Dietary specialization

Scales probably must be complemented by other kinds of food. Whitfield & Blaber (1978) provided the calorific value of scales from *Mugil cephalus* (= 2 cals mg^{-1}), a mugilid fish preyed on by the teraponid *Terapon jarbua*. Fish scales contain 40–85% protein (van Oosten 1957), but data on their nutritive value are entirely lacking. The mucus covering scales may be an important energy source for scale-eaters, as suggested for the cleaning labrid, *Labroides phthirophagus* (Gorlick 1980). Fish mucus is rich in protein (Wessler & Werner 1957) and lipids (Lewis 1970). Probably no scale-eating characoid feeds exclusively on scales, a possible exception being *Roeboexodon* (Roberts 1970, Knöppel 1972). *Roeboides* also ingests arthropods and fishes (Zaret & Rand 1971), and *Probolodus*, *Bryconexodon* and even *Exodon* take insects and plant material, including leaves, seeds, and algae (Roberts 1970, Knöppel 1972, Sazima 1977, 1980. Géry 1980, Sazima & Machado 1982). *Catoprion* eats insects and plant material, but has a distinct preference for scales when adult (Vieira & Géry 1979). Such a shift from a diversified diet to an almost entirely lepidophagous one as the fish grows, has also been noted for species of *Roeboides* (Roberts 1970, Sazima 1980, Sazima & Machado 1982).

The capacity of scale-eating characoids to take food items other than scales may be important. I examined a seemingly healthy adult *Probolodus heterostomus* with most of its teeth worn out and some in the process of replacement, its stomach containing only plant material and insects. Other scale-eating fishes also take alternative food items (Marlier & Leleup 1954, Greenwood 1965, Hoese 1966, Fryer & Iles 1972, Major 1973, Sazima & Uieda 1980, Whitfield & Blaber 1978). A few species, such as the cichlids *Perissodus straeleni* and *Corematodus* spp., and the triacanthodid *Tydemania navigatoris*, seem to subsist solely on scales (Trewavas 1947, Marlier & Leleup 1954, Fryer & Iles 1972, Mok 1978).

Scale-eating appears to be a size-limited behavior. Some scale-eating fishes, such as the carangid genus *Oligoplites*, cease lepidophagy, concomitantly with changes in dentition, as they grow (Major 1973, Smith-Vaniz & Staiger 1973, Sazima & Uieda 1980). Moreover, several African cichlids increasingly complement their diet with small fishes (Marlier & Leleup 1954, Fryer & Iles 1972). Perhaps it is no coincidence that scale-eaters rarely exceed the length of 200 mm and that most range near 120 mm (Roberts 1970, Fryer & Iles 1972, Mok 1978, Sazima 1980). Lower size limits occur too; *Probolodus* at length of about 19 mm takes scales from *Astyanax* of similar sizes (Sazima 1980).

In the stomach of lepidophagous fishes, the ingested scales become eroded and coalescent; in the intestine, they are reduced to a structureless pulp (Roberts 1970, Whitfield & Blaber 1978, Sazima & Machado 1982). The digestive transit time varies according to the species (Sazima 1980). As far as I know, no one to date has investigated the physiology of digestion in scale-eating fishes.

The proportion of other food items ingested, relative to scales, can provide a useful measure of the degree of specialization of a given scale-eating species. Stomach contents of 21 specimens of *Roeboides prognathus* taken at one site in the Rio Cuiabá, Mato Grosso State, were composed exclusively of scales, except for a few insects in five of them. In contrast, stomachs of 14 syntopical, comparably sized individuals of *R. bonariensis* contained insects in the same proportion as scales, with six of them having no scales at all. It seems that *R. bonariensis* has a more diversified diet, which agrees with its less modified jaws and teeth compared to *R. prognathus* (Sazima 1980). In the characid subfamily Acestrorhynchinae a trend towards piscivory was found in the most specialized species as defined by changes in dentition, cranial architecture, and body plan (Menezes 1969). A similar trend possibly occurs within the genus *Roeboides*, in which forms such as *R. bonariensis* seem to be the least specialized (near the ancestral, *Charax*-like stock?), whereas forms like *R. prognathus* appear the most specialized (Sazima 1980)

The evolution of scale-eating habits

Trophic origin

Several hypotheses have been proposed to explain the evolution of scale-eating in fishes, and all of them involve a trophic origin: scraping of epilithic or epiphytic algae, removal of epibionts or ectoparasites, modified and 'degenerated' forms of predation or piscivory, and necrophagy (Fryer et al. 1955, Greenwood 1965, Géry 1969, Fryer & Iles 1972, Whitfield & Blaber 1978, Whitfield 1979, Sazima & Uieda 1980, Sazima & Machado 1982).

Scale-eating African cichlids are said to have evolved both from herbivorous and carnivorous ancestors (Fryer et al. 1955, Greenwood 1965, Fryer & Iles 1972, Liem & Stewart 1976). Fryer et al. (1955) proposed an intermediate step toward lepidophagy, starting from algae-scraping behavior, through the grazing of epibionts on the bodies of other fishes. A 'modified piscivory' has also been proposed by some authors, although none has suggested what kinds of paths might have led to scale-eating. Of two closely related species of *Docimodus*, one has fin-eating habits whereas the other is a skin- and scale-eater (Eccles & Lewis 1976). The fact that some scale-eaters can be caught on lines baited with pieces of fish (Fryer & Iles 1972), however, does not contribute much to the argument for 'modified piscivory', inasmuch as feeding on small fishes is usual in certain size classes of scale-eaters.

Modified piscivory was also suggested by Géry (1964, 1969, 1972) for *Catoprion* and *Exodon*. For *Catoprion*, modified piscivory seems to be a reasonable mechanism, once accepting its phyletic affinity with the piranhas, Serrasalminae (Géry 1972, 1977). Most species of *Serrasalmus*, especially when young, feed mainly on fins or chunks of the body of other fishes. The species *S. elongatus* always appears to specialize on fin- and scale-eating (Roberts 1970, Goulding 1980). Some aspects of the hunting and scale-eating behaviors of *C. mento* are in many ways similar to those of some piranhas, and this seems to lend support to such an idea. However, as the evolution of particular habits is difficult to trace using exclusively behavioral data, phyletic studies are needed to help. At present the affinities between the Serrasalminae (piranhas) and the related Myleinae (pacus) are poorly understood and *Catoprion* may be a specialized offshoot of a group which evolved carnivorous as well as herbivorous habits.

The possible relationship between scale-eating, together with other 'mild' forms of mutilation, and the cleaning habit have been already pointed out (Losey 1978, 1979, DeMartini & Coyer 1981). The labrid *Labroides*, a well known cleaner, may forage on scales and mucus (Losey 1972b, Gorlick 1980), and the young of two kyphosid species engage in cleaning as well as scale-feeding (DeMartini & Coyer 1981). Scale-feeding in the carangids *Oligoplites* and *Scomberoides* probably is also related to occasional cleaning habits (Smith-Vaniz & Staiger 1973, Sazima & Uieda 1980), for one behavior might lead to the other — cleaning and scale- or mucus-eating are extremes on a behavioral continuum (DeMartini & Coyer 1981). For *Oligoplites*, necrophagy was proposed as an alternative (or complementary) origin for scale-eating (Sazima & Uieda 1980). Removal of scales from dead fishes has not been observed in lepidophages other than *O. saurus*. I suspect that some of the scales reported in *Arius cleptolepis* (Roberts 1978) might have been ingested in this manner, as I observed young individuals of other two ariids, *Netuma barba* and *Genidens genidens*, removing mucus and scales from dead fish (Sazima 1980).

Mucus-eating may have been the ancestral feeding behavior of the scale- and even blood-feeding trichomycterids. *Ochmacanthus reinhardtii* was reported feeding on this nutritious secretion (Roberts 1972) and *Branchioica bertonii*, aside from taking blood in the gill chambers of *Pseudoplatystoma* spp., were observed behaving as if they were rasping mucus from the body surface of these hosts (Machado & Sazima 1983). I have found specimens of *Pseudostegophilus maculatus* with their stomachs filled with mucus along with some characoid scales. Mucus seems to be of major importance in the diet of some ariids, blenniids, labrids, and even young cichlids (Hoese 1966, Hobson 1968, Losey 1972a, 1972b, 1978, Noakes 1979, Gorlick 1980, Sazima 1980).

Social origin

It seems logical to seek the origin of a given feeding habit in other seemingly related modes of feeding. Nevertheless, this will not necessarily be the case, for some specialized modes of feeding have almost certainly originated from other, qualitatively different behaviors (Eibl-Eibesfeldt 1970, Curio 1976). This seems to be the case for at least *Probolodus heterostomus* and *Exodon paradoxus*, for which I suggest here a primarily social origin for their specialized, scale-eating habits, i.e. intra- and interspecific aggression. As seen earlier, there is evidence that both species are related to the Tetragonopterinae, and it is certain that some *Astyanax* species occasionally consume scales. Let us examine the agonistic behavior of some tetragonopterines for the possible origins of scale-eating habits.

Many *Astyanax* species are inquisitive schooling fishes which feed opportunistically on almost anything edible. Aggressive encounters between individuals in the school are frequent and consist mainly of short chases and bites (sometimes accompanied by displays such as undulatory swimming). Occasionally in these conflicts a few scales are dislodged by the attacking fish, which may then ingest them. This behavior is frequent in *Astyanax bimaculatus*, *A. fasciatus* and *A. scabripinnis* and, indeed, appreciable amounts of *Astyanax* scales may be consumed by the two former species (Nomura 1975). Breder (1927) commented upon the aggressiveness of *A. ruberrimus* and noted that most individuals bear wounds or scars resulting from bites.

Aggressive behavior in *Astyanax* may intensify during and after feeding activities, seemingly because search for food increases the chance and the duration of the encounters between the members of the school. Such a correlation has previously been recognized in animals other than fish (Curio 1976). Moreover, tetragonopterine characins often congregate in schools of mixed species, or even genera (Lowe-McConnell 1975), and aggressive scale-eating behavior may sometimes be directed against fishes other than conspecifics. Were this to have repeatedly occurred in crowded, food-limited populations, seemingly ideal conditions for the initial evolution of scale-eating would have been created. It seems that from the 'mild' scale-eating practiced by some *Astyanax*, very few evolutionary steps are necessary to arrive at the specialized, habitual lepidophagy found in *Probolodus heterostomus* and *Exodon paradoxus*. Both species have general habits similar to those of some *Astyanax* species and are also quarrelsome towards conspecifics and other fish (Sazima 1980). The agonistic displays of *Probolodus* exhibited at the sudden approach of another fish in the school, as well as the unique 'twin spot' pattern of *Exodon*, probably have appeared as aggression-reducing devices during the evolution of lepidophagous habits, thereby reducing intraspecific scale-eating (Sazima 1980).

Two important conditions for the acquisition of lepidophagy via aggressive behavior as envisioned here are that agonistic encounters increase in feeding occasions and that attacks are occasionally directed towards individuals of other species in the mixed schools (both observed in *Astyanax*). Another point of importance is the propensity to ingest the scales dislodged during aggressive encounters, as some fishes do not make use of them, whereas *Astyanax* swallow the scales and digest them to a certain extent. The origin of scale-eating in *Probolodus* and *Exodon* via aggressive behavior in an *Astyanax*-like ancestor may presently be accepted as a probable mode of evolution of this feeding behavior for these two genera. Phyletic studies on *Probolodus*, *Exodon*, *Bryconexodon*, *Astyanax* and other seemingly related characids may provide an additional test for this interpretation. (Of course, there still remains the alternative of an exclusively trophic origin.)

Intraspecific aggression might also provide an alternative explanation for the origin of scale-eating in *Catoprion mento*, a serrasalmine. Although this may seem unduly speculative, it seems relevant in this context to quote Greenwood (1965) who, in discussing the evolution of scale-eating habits of the African cichlid *Haplochromis welcommei*, wrote: 'If the derivation of the lepidophagous species from algae-grazing ancestors be accepted, there is still one major step unexplained: what stimulus led the proto-lepidophages to seek food from the surface of other fishes?' One possible answer is aggressive behavior. Most cichlids exhibit aggressive, territorial behavior (Eibl-Eibesfeldt 1970, Fryer & Iles 1972), and some of their agonistic encounters might result in biting and scale-removal. I have observed captive individuals of two unidentified species of herbivorous myleine characids chasing, biting and swallowing the dislodged scales of conspecifics. A herbivorous species adapted to graze on hard substrates, such as a number of African 'mbuna' species, could very well dislodge and ingest some scales in attack situations. From this origin, scale-removal and ingestion may be perfected to constitute a habitual mode of feeding dislinked from its original social context.

Table 1. Behavioral characteristics, feeding habits and suggested origin of lepidophagy, in various scale-eating fishes.[a]

Species[b]	Social habits	Hunting tactic	Scale-removal behavior	Main food items	Suggested origin for scale-eating
CHARACIDAE					
Roeboides prognathus (plain color, young translucent)	Solitary, rarely in small groups	Ambushes or stalks	Strikes at the flank of prey, mainly with mouth closed[c]	Scales, insects, fishes	Ingestion of scales during unsuccessful attack to prey; Picking up epibionts
Exodon paradoxus (twin spot pattern)	Schooling (in homotypic groups)	Rushes toward prey, in group	Strikes at the flank of prey, with mouth closed, or bites[c]	Scales, insects	Intra- and interspecific aggressive behavior; Opportunistic foraging on various substrates
Probolodus heterostomus (aggressive mimic)	Schooling (mainly in mixed groups, together with prey)	Approaches under disguise	Strikes from behind, with mouth open and bites[d]	Scales, insects, plant material	As above
Catoprion mento (plain color)	Solitary, rarely in twos	Stalks, or ambushes, frequently with use of plant cover	Strikes at the flank of prey almost in a right angle and bites[d]	As above	Modified predation; Aggressive behavior
CICHLIDAE					
Corematodus shiranus (aggressive mimic)	Solitary, but joins prey's schools	Approaches under disguise	Closes its jaws over the tail of prey[e] (inferred)	Scales only	Modified predation
Perissodus straeleni (aggressive mimic)	As above	As above	Pushes its mouth against the back of prey[f]	Scales, fishes	Modified predation; Grazing on epibionts; Aggressive behavior
GARANGIDAE (young only)					
Oligoplites saurus (plain color)	Solitary, or in small groups	Swims together with prey, sometimes under disguise	Strikes with the side of mandible parallel to the flank of prey[g]	Scales, invertebrates, fishes	Opportunistic foraging on other fishes ('cleaning'); Necrophagy
Oligoplites palometa (barred pattern)	Solitary	Stalks, frequently with use of cover	As above	As above	As above
TERAPONIDAE					
Terapon jarbua (striped pattern)	Schooling	Schools and lunges at prey	Presses open jaws against the prey's body[h]	Scales, invertebrates, fishes, plant material	Modified predation; Modified parasite removal
BLENNIIDAE					
Plagiotremus azalea and *P. laudandus* (aggressive mimics)	Solitary, but may join schools of other, similar fishes	Ambushes or stalks, frequently under cover or may disguise among other, similar fishes	Strikes from behind and bites at the prey's fins	Skin, scales, mucus	Opportunistic foraging on other fishes

[a] Sources: Trewavas 1947, Marlier & Leleup 1954, Géry 1964, 1969, Hobson 1968, Roberts 1970, Fryer & Iles 1972, Losey 1972a, Sazima 1977, 1980, Brichard 1978, Whitfield & Blaber 1978, Whitfield 1979, Sazima & Uieda 1980, Sazima & Machado 1982, present paper.

[b] The geographical distribution of these species is: Characidae – S. America (freshwater); Cichlidae – Africa (freshwater); Carangidae – W. Atlantic; Teraponidae – Indo-Pacific; Blenniidae – Pacific.

[c] Teeth conical or mamilla-like, some external and pointed almost straight ahead.

[d] Teeth tricuspid, with median cusp greatly developed and pointed outwards.

[e] Teeth small, oblique crowned pointed, and arranged in file-like series.

[f] Broad, laminar and strongly recurved teeth.

[g] Outer dentary teeth hooked outwards, with spatulate tips.

[h] Teeth conical, the outer series larger than the inner, the latter forming a broad band.

I suggest that scale-eating habits may be initiated, or facilitated, by two types of causes: trophic or social. These may be complementary, rather than mutually exclusive, although their relative importance can — and probably does — vary from situation to situation. The behavioral link provided by aggression may help explain some of the steps needed for the evolution from a given, more 'orthodox' feeding habit, to a predominantly lepidophagous one. This may be especially true for such new, opportunistic groups, as were the characoids over the course of their adaptive radiation. Since scale-eating arose in diverse, independent evolutionary lines among the various fish groups, many behavioral alternatives have certainly been involved in the establishment of this feeding habit (Table 1).

Acknowledgments

I have benefited greatly from the discussion and suggestions of N.A. Menezes, H.A. Britski, W.W. Benson, C.G. Froehlich, P.V. Loiselle, D.L. Kramer, T.R. Roberts, F.A. Machado and T.M. Lewinsohn. An earlier version was reviewed by T.M. Zaret, W.W. Benson, D.L. Kramer and T.R. Roberts, who transformed my English writing into a seemingly acceptable product. Other contributions, either in field or laboratory, or through their generous hospitality include: M. Sazima, J. de Lima Figueiredo, W.C.A. Bokermann, E.P. and U. Caramaschi, M. Vieira da Silva and family, V.S. and W. Uieda, A Machado, J.B. da Silva, J. de Oliveira, E.F. Nonato, C. Aranha, K. Dobat, and K. Rohde. The final part of the study was supported in part by a grant from the Conselho Nacional de Desenvolvimento Científico e Tecnológico.

References cited

Baskin, J.N., T.A. Zaret & F. Mago-Leccia. 1980. Feeding of reportedly parasitic catfishes (Trichomycteridae and Cetopsidae) in the Rio Portuguesa Basin, Venezuela. Biotropica 12: 182–186.

Breder, C.M., Jr. 1927. The fishes of the Rio Chucunaque drainage, eastern Panama. Bull. Amer. Mus. Nat. Hist. 57: 91.176.

Brichard, P. 1978. Fishes of Lake Tanganyika. T.F.H. Publ., Neptune City. 448 pp.

Curio, E. 1976. The ethology of predation. Springer-Verlag, Berlin. 249 pp.

DeMartini, E.E. & J.A. Coyer. 1981. Cleaning and scale-eating in juveniles of the kyphosid fishes, *Hermosilla azurea* and *Girella nigricans*. Copeia 1981: 785–789.

Eccles, D.H. & D.S.C. Lewis. 1976. A revision of the genus *Docimodus* Boulenger (Pisces: Cichlidae), a group of fishes with unusual feeding habits from Lake Malawi. Zool. J. Linn. Soc. 58: 165–172.

Eibl-Eibesfeldt, I. 1970. Ethology, the biology of behavior. Holt, Rinehart & Winston, New York. 540 pp.

Fryer, G., P.H. Greenwood & F. Trewavas. 1955. Scale-eating habits of African cichlid fishes. Nature 175: 1089–1090.

Fryer, G. & T.D. Iles. 1972. The cichlid fishes of the Great Lakes of Arica, their biology and evolution. Oliver & Boyd, Edinburgh. 641 pp.

Géry, J. 1959. Contribution à l'étude des poissons Characoides (Ostariophysi) (II.) *Roeboexodon* gen.n. de Guyane, redescription de *R. guyanensis* (Puyo, 1948) et relations probables avec les formes voisines. Bull. Mus. Nat. Hist. Nat. 31: 403–409.

Géry, J. 1964. Poissons characoides nouveaux ou non signalés de L'ilha do Bananal. Brésil. Vie et Milieu supl. 17: 447–471.

Géry, J. 1966. Endemic characoid fishes from the upper Rio Cauca at Cali, Colombia, Ichthyologica, The Aquarium Journal (January): 13–24.

Géry, J. 1969. The fresh-water fishes of South America. pp. 828–848. *In*: E.J. Fittkau et al. (ed.) Biogeography and Ecology in South America, vol. 2, Dr. W. Junk Publishers, The Hague.

Géry, J. 1972. Poisson characoides des Guyanes. I. Généralités. II. Famille des Serrasalmidae. Zool. Verhand. 1–241.

Géry, J. 1977. Characoids of the world. T.F.H. Publ., Neptune City. 672 pp.

Géry, J. 1980. Un nouveau poisson characoide occupant la niche des mangeurs d'écailles dans le haut rio Tapajoz, Brésil: *Bryconexodon juruenae* n.g.sp. Rev. fr. Aquariol. 7: 1–8.

Gorlick, D.L. 1980. Ingestion of host fish surface mucus by the Hawaiian cleaning wrasse, *Labroides phthirophagus* (Labridae), and its effects on host species preference. Copeia 1980: 863–868.

Goulding, M. 1980. The fishes and the forest: explorations in Amazonian natural history. Univ. California Press, Berkely. 280 pp.

Greenwood, P.H. 1965. Two new species of *Haplochromis* (Pisces, Cichlidae) from Lake Victoria. Ann. Mag. Nat. Hist. Ser. 13: 303–318.

Hobson, E.S. 1968. Predatory behavior of some shore fishes in the Gulf of California. Res. Rep. U. S. Fish. Wildl. Serv. 73: 1–92.

Hoese, H.D. 1966. Ectoparasitism by juvenile sea catfish, *Galeichthys felis*. Copeia 1966: 880–881.

Knöppel, H.A. 1972. Zur Nahrung tropischer Süsswasserfische aus Südamerika. Amazoniana 3: 231–246.

Kramer, D.L. 1978. Reproductive seasonality in the fishes of a tropical stream. Ecology 59: 976–985.

Lewis, R.W. 1970. Fish cutaneous mucus: a new source of skin surface lipids. Lipids 5: 947–949.

Liem, K.F. & J.W.M. Osse. 1975. Biological versatility, evolution and food resource exploitation in African cichlid fishes. Amer. Zool. 15: 427–454.

Liem, K.F. & D.J. Stewart. 1976. Evolution of the scale-eating cichlid fishes of Lake Tanganyika: a generic revision with a description of a new species. Bull. Mus. Comp. Zool. 147: 319–350.

Losey, G.S. 1972a. Predation protection in the poison-fang blenny, *Meiacanthus atrodorsalis*, and its mimics, *Ecsenius bicolor* and *Runula laudanus* (Blenniidae). Pacif. Sci. 26: 129–139.

Losey, G.S., Jr. 1972b. The ecological importance of cleaning symbiosis. Copeia 1972: 820–833.

Losey, G.S., Jr. 1978. The symbiotic behavior of fishes. pp. 1–31. *In*: D.I. Mostofsky (ed.) The Behavior of Fish and Other Aquatic Animals, Academic Press, New York.

Losey, G.S., Jr. 1979. Fish cleaning symbiosis: proximate causes of host behaviour. Anim. Behav. 27: 669–685.

Lowe-McConnell, R.H. 1964. The fishes of the Rupununi savanna district of British Guiana, South America. Pt. 1. Ecological grouping of fish species and effects of the seasonal cycle on fish. J. Linn. Soc. (Zool.) 45: 103–144.

Lowe-McConnell, R.H. 1975. Fish communities in tropical freshwaters. Longman, London. 337 pp.

Machado, F.A. & I. Sazima. 1983. Feeding behavior of the blood-feeding fish *Branchioica bertonii* (Siluriformes, Trichomycteridae). Ci. e Cult. 34: (in press). (In Portuguese).

Major, P.F. 1973. Scale-feeding behavior of the leatherjacket, *Scomberoides lysan* and two species of the genus *Oligoplites* (Pisces: Carangidae). Copeia 1973: 151–154.

Markl, H. 1972. Aggression und Beuteverhalten bei Piranhas (Serrasalminae, Characidae). Z. Tierpsychol. 30: 190–216.

Marlier, G. & N. Leleup. 1954. A curious ecological 'niche' among the fishes of Lake Tanganyika. Nature 174: 935–936.

Menezes, N.A. 1969. Systematics and evolution of the tribe Acestrorhynchini (Pisces, Characidae). Arq. Zool. S. Paulo 18: 1–150.

Mok, H.K. 1978. Scale-feeding in *Tydemania navigatoris* (Pisces: Triacanthodidae). Copeia 1978: 338–340.

Noakes, D.L.G. 1979. Parent touching behavior by young fishes: incidence, function and causation. Env. Biol. Fish. 4: 389–400.

Nomura, H. 1975. The food of three species of *Astyanax* Baird & Girard, 1854 (Osteichthyes, Characidae) of the Mogi Guaçu River, SP. Revta.bras. Biol. 35: 595–614. (In Portuguese).

Roberts, T.R. 1970. Scale-eating American characoid fishes, with special reference to *Probolodus heterostomus*. Proc. Calif. Acad. Sci. 38: 383–390.

Roberts, T.R. 1972. Ecology of fishes in the Amazon and Congo basins. Bull. Mus. Comp. Zool. 143: 117–147.

Roberts, T.R. 1978. An ichthyological survey of the Fly River in Papua New Guinea with descriptions of new species. Smithsonian Contr. Zool. 281: 1–72.

Sazima, I. 1977. Possible case of aggressive mimicry in a neotropical scale-eating fish. Nature 270: 510–512.

Sazima, I. 1980. A comparative study of some scale-eating fishes (Osteichthyes). Ph. D. Thesis, Universidade de São Paulo, São Paulo. 264 pp. (In Portuguese).

Sazima, I. & F.A. Machado. 1982. Habits and behavior of *Roeboides prognathus*, a scale-eating fish (Osteichthyes, Characoidei). Bolm. Zool. Univ. S. Paulo 7: 35–56. (In Portuguese).

Sazima, I. & V.S. Uieda. 1980. Scale-eating behavior in *Oligoplites saurus* and records of scale-eating in *O. palometa and O. saliens* (Pisces, Carangidae). Revta. bras. Biol. 40: 701–710. (In Portuguese).

Smith-Vaniz, W.F. & J.C. Staiger. 1973. Comparative revision of *Scomberoides*, *Oligoplites*, *Parona* and *Hypacanthus* with comments on the phylogenetic position of *Campogramma* (Pisces: Carangidae). Proc. Calif. Acad. Sci. 39: 185–256.

Trewavas, E. 1947. An example of 'mimicry' in fishes. Nature 160: 120.

Van Oosten, J. 1957. The skin and scales. pp. 207–224. *In*: M.E. Brown (ed.) The Physiology of Fishes, Academic Press, New York.

Vleira, I. & J. Géry. 1979. Differential growth and nutrition in *Catoprion mento* (Characoidei). Scale-eating fish of Amazônia. Acta Amazonica 9: 143–146. (In Portuguese).

Wessler, E. & I. Werner. 1957. On the chemical composition of some mucous substances of fish. Acta Chem. Scand. 2: 1240–1247.

Whitfield, A.K. 1979. Field observations on the lepidophagous teleost *Terapon jarbua* (Fôrskal). Env. Biol. Fish. 4: 171–172.

Whitfield, A.K. & S.J.M. Blaber. 1978. Scale-eating habits of the marine teleost *Terapon jarbua* (Fôrskal). J. Fish. Biol. 12: 61–70.

Zaret, T.M. & A.S. Rand. 1971. Competition in tropical stream fishes: support for the competitive exclusion principle. Ecology 52: 336–342.

Originally published in Env. Biol. Fish. 9: 87–101

Grazing responses of tropical freshwater fishes to different scales of variation in their food

Mary E. Power
Division of Environmental Studies, University of California, Davis, California 95616, U.S.A.

Keywords: Algivory, Habitat choice, Loricariids, Neotropical streams, Periphyton, Seasonality, Ideal free distribution

Synopsis

Grazing fishes in neotropical streams confront variation in their attached algal food that ranges in scale from differences in quality among algal cells to differences in the primary productivity of habitats available to the fishes. Fishes may respond to this variation on some scales but not others. For example, loricariid catfish in a Panamanian stream tracked variation in algal productivity among pool habitats very closely. In sunny pools where algae grew about seven times faster than in shaded pools, loricariids were six to seven times denser. Consequently, growth rates of pre-reproductive *Ancistrus spinosus* (the most common species in pools) were similar in pools of different canopies, corresponding to predictions from the 'ideal free distribution' hypothesis. But on a smaller scale, within pools, avoidance of avian and terrestrial predators outweighed foraging considerations. Larger species and size classes avoided water shallower than 20 cm, where (as a result) the only standing crops of attached algae large enough to be measurable by scraping occurred. During the dry season when food was most limiting, loricariids overlapped more in their substrate use as different species sought cover in common refuges such as logs and root tangles in pools. Seasonal variation in growth rates of pool-dwelling loricariids reflect these constraints.

Introduction

Fishes foraging in neotropical streams confront considerable temporal and spatial variation in their food. How do they respond to this variation, and to what extent do their feeding responses account for their numbers and distributions in streams? In this paper, I address these questions for fishes that graze periphyton, or attached algae, from stream substrates. Here I consider grazers to be animals that eat foods that are small, sessile and fine-grained (MacArthur & Levins 1964) in their distribution relative to their consumer. Distinctions between grazing and browsing herbivorous fishes made by Hiatt & Strasbourg (1960) and Keenleyside (1979) are followed here: grazers crop their food so closely that they often ingest bits of the substrate, while browsers nip off vegetation further from the substrate and do not ingest appreciable amounts of it. Following Wetzel (1975, p. 390), I use periphyton to refer to microfloral growth on various aquatic substrates. This usage overlaps with the term 'aufwuchs' as used by some authors (Ruttner 1952, Fryer & Iles 1972).

Variation in the availability of periphyton for grazing stream fishes occurs on small, intermediate and large spatial scales. On a small scale, there are differences in the nutritional value of organisms within the periphyton community. On an intermediate scale, within pool or riffle habitats of streams,

Thomas M. Zaret (ed.), Evolutionary ecology of neotropical freshwater fishes. ISBN 90 6193 823 6

algivorous fishes encounter both continuous and discontinuous spatial variation in their food. Gradients in periphyton standing crops and community compositions occur with depth. At a given depth, periphyton varies on different substrates such as logs, cobbles, sand or mud (Blum 1956, Round 1964, Hynes 1970, Whitton 1975). Grazers themselves create mosaics of patches, even on initially homogeneous substrates, by leaving different sites in various stages of recovery.

On a large scale, differences in the density of forest canopy over a stream cause large differences in the primary production of periphyton below. Variation in streams on a yet larger scale occurs from the headwaters to the river mouths (Cummins 1977, Vannote et al. 1980), but this scale will not be considered here. Heterogeneity from all sources affects the rates and efficiencies with which algivorous fishes harvest foods.

In this paper, I discuss adaptations and constraints of grazing fishes that affect their responses to small, intermediate and large scale variation in periphyton food. For large and intermediate scales, I emphasize observations on armored catfish (Loricariidae) that graze periphyton in a Panamanian stream.

Small scale variation

Nutritional quality of periphyton

Periphyton communities in streams commonly include diatoms and encrusting or filamentous bluegreen, green and red algae (Round 1965, Hynes 1970). Our knowledge of the nutritional qualities of freshwater algae for fishes is still sketchy, but considerable variation has been uncovered in studies of algal grazing by invertebrates (Porter 1977).

Herbivores are often limited by the quality, in particular the protein content, of their food (Westoby 1974, White 1978). The protein content of thirteen species of freshwater algae ranged from 10 to 46% of the dry weight, with the highest protein concentrations occurring in *Euglena* and three bluegreen algae, *Anabaena*, *Microcystis* and *Aphanizomenon* (Boyd 1973). Interestingly, the protein content of these algae (42 to 46%) roughly coincides with optimal protein concentrations for catfish juveniles grown on artificial diets. Dupree & Sneed (1966) grew channel catfish, *Ictalurus punctatus*, on diets of constant caloric content, but with protein levels ranging from 12 to 52%. Catfish gained weight faster as diet protein increased until it was about 40%, after which weight gain decreased. Digestion of protein entails higher metabolic costs than digestion of fats or carbohydrates, and the decreased growth on extremely rich protein diets was presumably due to increased metabolic costs in processing these foods (discussed in Boyd & Goodyear 1971).

But high protein content in bluegreen algae may be offset by toxins (Gentile 1971, Willoughby 1977) or by cell walls that resist digestion (Fish 1951, Porter 1977, but see Moriarty & Moriarty 1973). Dussault & Kramer (1981) reared guppies, *Poecilia reticulata*, on pure diets of the green alga *Chlorococcum* and the filamentous bluegreen, *Oedogonium*. The fish grew and matured on the first diet, but not on the second. Bluegreen algae have also been found inferior to other types of algae as foods for a number of invertebrates (Calow 1975, Porter 1977).

Algivorous fishes can break down cell walls in three ways: mechanical grinding, acid lysis or cellulase enzymes derived from gut microflora. Mullets (*Mugil* spp.) and some herbivorous reef fishes have thick-walled, gizzard-like portions of their stomachs where they use ingested mineral particles to grind cell walls of algae and bacteria (Hiatt & Strasbourg 1960, Odum 1970, Payne 1978, Ogden & Lobel 1978). *Tilapia nilotica* and *Haplochromis nigripinnis*, cichlids that feed on bluegreen algae, can lower the pH of their stomachs to about 1.4, at which point bluegreen cell walls are completely lysed (Moriarty & Moriarty 1973). Acid secretion in the stomachs of these cichlids, however, begins only when ingestion starts at dawn. Stomach pH is not at its minimum until late morning, so four to six hours of feeding elapse before assimilation of the ingested algae is up to its maximum efficiency of 70 to 80% (Moriarty & Moriarty 1973). Cellulase enzymes which lyse bluegreen cell walls have never been found in vertebrates unless associated with microbes or invertebrates in the gut (Prejs & Blaszczyk 1977). Cellulase activity, however, has been found in the guts of several fishes (Stickney & Shumway 1974, Prejs & Blaszczyk 1977, Niederholzer & Hofer 1977). Channel catfish, *Ictalurus punctatus*, had cellulase activities in their guts until they were exposed to streptomycin, indicating that enzymes were derived from gut microbes (Stickney & Shumway 1974). A congener, *I. nebulosus*, perhaps assisted by gut microflora, was able to digest and assimilate the bluegreen alga *Anabaena flos-aquae* with 67.5% efficiency, while the green alga *Spirogyra* was only 23.7% assimilated (Gunn et al. 1977). The high assimilation of the protein rich bluegreen alga suggests that bluegreens are potentially important foods for these catfish, as they make up large portions of the gut contents of catfish from natural habitats (Gunn et al. 1977).

Striped mullet, *Mugil cephalus*, have long coiled intestines up to five times the length of their bodies, which contain flagellates that may digest cellulose (Odum 1970). Relatively long intestines are also found in loricariids. In loricariid feces, intact bluegreen and green algal filaments that appear viable are common, while nearly all diatom frustrules are empty, suggesting that the latter are more digestible for loricariids (personal observations).

In summary, nutritional quality of algae for algivorous fishes depends both on properties of the algae (including the medium in which they have grown, Spoehr & Milner 1949, Gerloff & Skoog 1954), and on adaptations of the fishes. Sufficient variation in nutritional values of components of periphyton exists to suggest that selective ingestion might be advantageous for algae-grazing fishes. On the other hand, algivores like other herbivores might depend on different components of their diets for different nutrients (Westoby 1974) or need to restrict their intake of different toxins in particular foods (Freeland & Janzen 1974), and therefore might require a mixed diet.

Diet choice

The advantage for algae-grazing fishes in discriminating among the foods they ingest is an issue separate from their ability to do so. Some herbivorous damselfishes browse macroalgae from substrates on marine reefs selectively (Lassuy 1980) while others feed unselectively (Montgomery 1980). But selective ingestion of items seems less feasible for fishes grazing microalgae only 10's or 100's of microns in diameter, given that the fishes' mouths are several millimeters or centimeters wide. One of the more delicately browsing algivorous fishes, *Poecilia reticulata*, was observed by Dussault & Kramer (1981), who used cinematography. With each bite, the guppies ate $3\,mm^2$, an area equivalent to about 10^4 diatoms of typical size lain flat in a layer one cell thick.

Selective ingestion of organisms from algal mats is probably more difficult for grazers than for browsers. Mouths of many bottom grazing fishes have thick, fleshy, often suctorial lips, unsuited for selective nipping. Such mouths occur in loricariid catfish, prochilodontid characins (Roberts 1973), grazing cyprinids such as the African *Labeo*, and *Gyrinocheilus*, the mountain carp of Borneo and Thailand (Hora 1933). In addition, the morphology of loricarids is such that they cannot see the substrate on which they are grazing (Fig. 1).

The ability of algivorous fishes to select certain components of periphyton must depend on the spatial distribution of these components. *Haplochromis guentheri*, an African 'mbuna' cichlid from Lake Malawi, selectively nips off long filaments of green and bluegreen algae which protrude from the algal felt on rock substrates (Fryer 1959). Snails, *Planorbis contortus*, feeding with radulas selectively ingest bacteria from detritus on stones (Calow 1974, 1975). The ability of the snails to select bacteria may depend both on clumping by the bacteria and on chemical sensing of these aggregates by the snails (Calow 1974). Physical environmental factors may influence clumping by algae. In culture, a bluegreen, *Anabaena cylindrica*, develops gelatinous walls which cause filaments to clump when the alga is incubated in water with sodium: calcium ratios similar to those in a Kenyan soda lake (Fryer & Iles 1972). Perhaps grazing fishes could more feasibly select or reject this alga in such environments.

If attached algal communities are stratified,

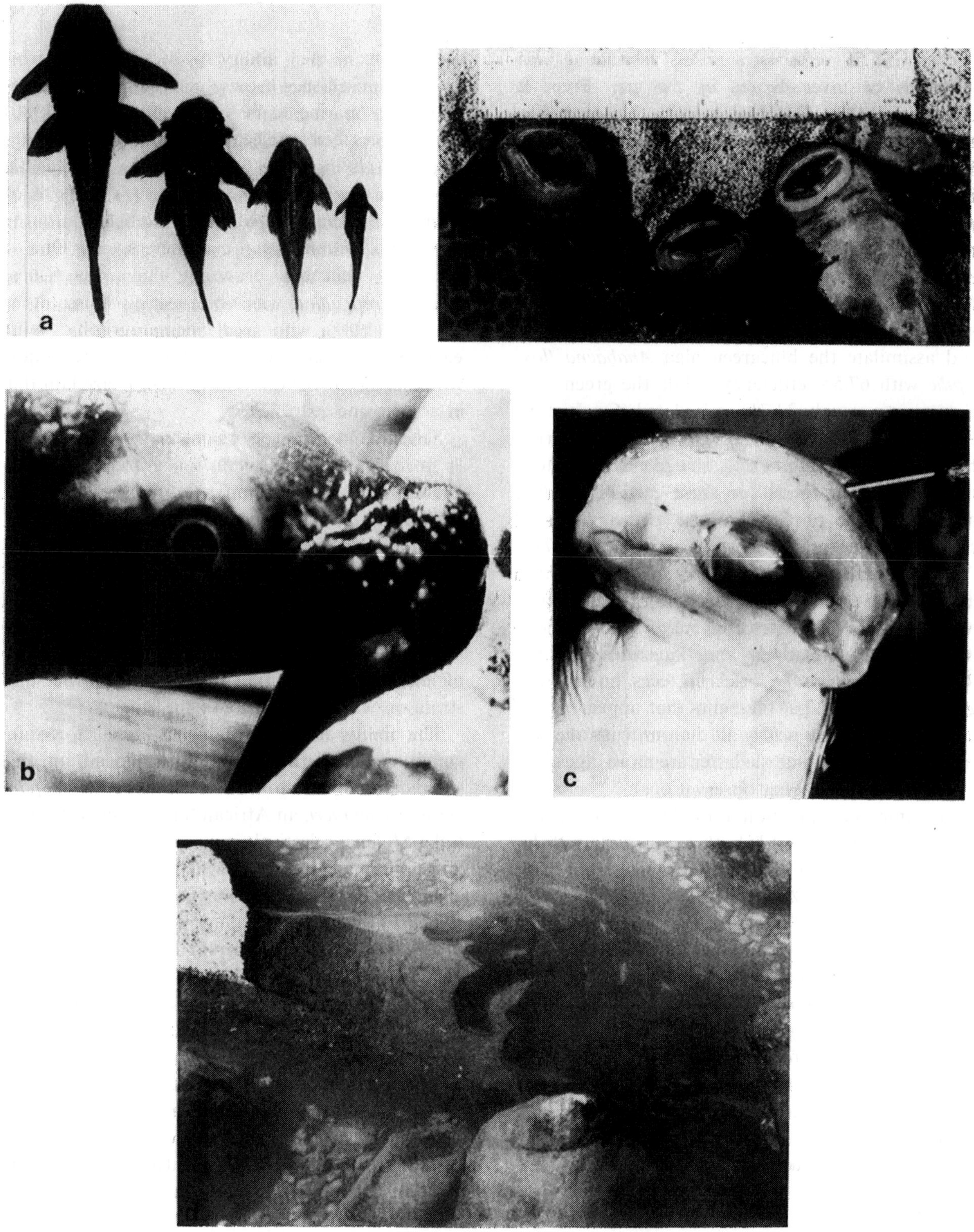

Fig. 1. a – Dorsal view and ventral view of mouths of the four loricariids that graze periphyton in the Rio Frijoles of Central Panama; from left to right, they are *Hypostomus plecostomus*, *Ancistrus spinosus*, *Chaetostoma fischeri*, and *Rineloricaria uracantha*. b – Head (in approximate feeding position), c – mouth and d – grazing marks of *Labeo coubie* in the Black Volta River, Bui, Ghana (courtesy of D.L. Kramer).

the depth at which fish penetrate while feeding ('grazing severity', sensu Alcock 1964) would affect the composition of their diet. Gastropods grazing marine microalgae selectively remove those diatoms which are more loosely adhered and superficially positioned in the algal mat (Nicotri 1977). Conceivably, if grazing severity were increased during periods of relative food shortage, grazer diets might include a wider range of microflora. This would be consistent with predictions from the foraging theory developed for animals that consume coarse-grained prey (MacArthur & Pianka 1966, Emlen 1966, Schoener 1971).

Because of the information just presented it is likely that a grazing fish's diet depends largely on its choice of feeding sites, because the composition of attached algal communities changes with position and substrate in streams. Some examples of variation of periphyton on this intermediate spatial scale are considered next.

Intermediate scale variation

Different physical and chemical conditions favor different algae. Therefore, the compositions of periphyton communities in streams vary with depth, exposure to current and substrate (Blum 1956, Round 1964, Hynes 1970, Whitton 1975). But to algivorous fishes, the physical features that govern their access to periphyton in various microhabitats may be more important than periphyton composition. These factors include substrate rugosity, sedimentation or currents that affect energy expended by fish while grazing, and cover available from predators at or near the grazing site. (By cover, I mean any factor that reduces vulnerability to predators, for example, physical shelter or sufficient depth of water to impede terrestrial or avian predators.)

Depth gradients

The richest standing crops of periphyton often occur at the shallow margins of streams and in shallow riffles, where exposure to light and nutrient fluxes are maximal, and where habitats first intercept nutrients washed in from the land. But algivorous fishes, particularly after they have grown large, do not occur in higher densities in shallow water, where food availability is presumably greater. Larger fishes generally stay in deeper water (Hellier 1962, Bowen 1979, De Silva & Silva 1979, Power 1983a), probably to avoid avian and terrestrial predators. For example, algae-grazing catfish (Loricariidae) in the Rio Frijoles of Central Panama are vulnerable to herons and kingfishers that commonly fish in water less than 20 cm deep. After loricariids grow too large to hide under cobbles in riffles, they almost never occur in water shallower than 20 cm, although they do move into shallower water by night (Power 1983a).

To quantify periphyton availability as a function of depth, I scraped samples from cobbles collected during March, near the end of the dry season. In water less than 10 cm deep, the median standing crop was 0.68 mg ash free dry weight (afdw) cm^{-2} (range = 0.27 to 1.50, N = 17). In water 10 to 20 cm deep, the median standing crop was 0.18 mg afdw cm^{-2} (range = 0 to 0.26, N = 9). In water deeper than 20 cm, I could not scrape sufficient periphyton from anywhere in the stream channel within a 3 km reach to be measurable within the errors of my technique. Larger loricariids were clearly food-limited during the dry season, when they stopped growing and sometimes lost weight (Power 1981, and see below). But they did not venture into water shallower than 20 cm to graze, even during the end of the dry season when food was most limiting in deeper water.

Substrates

Roberts (1972) has pointed out that algivorous fishes, whose gut contents appear indistinguishable, may still be feeding from different substrates or in different microhabitats. To study the distribution of Rio Frijoles loricariids over various substrates and depth intervals, I mapped a 3 km reach of the stream. During the dry season, this reach was surveyed with a hand level. In the upstream kilometer, cross-stream transects were made at roughly 9 m intervals, or wherever a major change in the stream's path or depth occurred. A meter tape was stretched across the stream and sufficient depth measurements made to construct a bathymetric map with 10 cm contour intervals. The stream banks were also mapped up to contour lines 40 cm above the dry season water level so that the map could be used during the higher stream stages of the rainy season. The lower 2 km were mapped in less detail, by surveying up the thalweg (i.e. deepest part) of the stream and noting only the positions of the water's edge and the points 40 cm above dry season water level. After a base map was drawn from these measurements, details were sketched in from spot measurements made in topographically complex areas. Throughout the 3 km reach, substrate type was also mapped. Within the mapped 3 km reach, 150 quadrats, 1 m^2 in area, were nailed into the streambed and located with respect to nearby permanent vegetation. Loricariids in these quadrats were censused by day and night by snorkelling (Power 1981), and substrates on which the various species occurred were noted.

Mud, sand, pebbles, cobbles, leaf mats, wood, submerged grasses, consolidated clay banks and bedrock platforms were available as substrates to loricariids in the Rio Frijoles. Bedrock and wood substrates were typical in pools along the outer walls of meander bends where the stream had been turned by bedrock formations or trees. Pebbles and cobbles were the predominant substrates in riffles. Leaf mats and mud accumulated in the deeper, slower water during the dry season, but were flushed away during the rainy season. The relative area of substrates in the 3 km reach of the Rio Frijoles during the dry and the rainy seasons are shown in Figure 2 along with the average proportions of loricariids of each of the four species sighted on a given substrate. All size classes for a given species have been pooled. Substrate use by different loricariid size classes within a species will be discussed elsewhere (Power 1981 and in preparation).

The most common pool-dwelling loricariid in the Rio Frijoles, *Ancistrus*, strongly preferred wood as a substrate in both the dry and the rainy season (see Fig. 2). Their high proportional occurrence on wood relative to its availability resulted in high densities of *Ancistrus* in root tangles and deadfalls in the stream. This species also commonly grazed on bedrock and clay substrates in pools. Algae were relatively accessible to wide-mouthed grazers such as *Ancistrus* on these flat substrates, which also offered solid support, energetically important to loricariids whose adaptations for benthic life include reduced swim bladders. *Ancistrus* commonly rested, but did not graze, on sandy floors of deep pools. In the dry season, their apparent preference for sand, coupled with increased avoidance of pebbles, resulted from the concentration of these fish in the deeper portions of pools as the water level dropped.

Hypostomus (= *Plecostomus*) rested by day in pools, under ledges or on the sandy bottoms. By night, they ventured out to graze cobble substrates in the deeper portions of riffles. In the dry season,

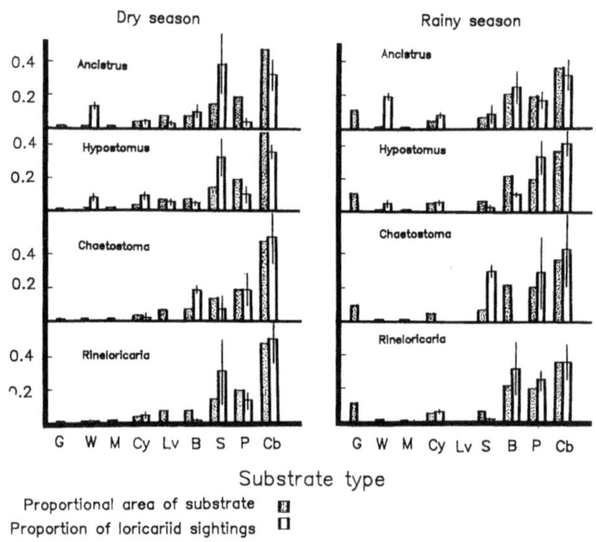

Fig. 2. Substrate use by loricariids and availability of substrate in the Rio Frijoles during dry and rainy seasons. Stippled bars show the proportion of stream channel area under various substrates in a 3 km reach of the stream. Substrate types, in order of their dry season abundance, are: G = submerged grasses, W = wood (large logs and roots), M = mud, Cy = consolidated clay, Lv = leaves and twigs, B = bedrock, S = sand, P = pebbles, Cb = cobbles. Open bars show the average proportions of all loricariids of a given species sighted during censuses that were on a given substrate. Census series, consisting of two day and two night censuses of over 150 1 m^2 quadrats distributed over the 3 km reach, were carried out in the dry season in March and April, and in the rainy season in August and October (see Power 1981 for methods). Bars show the range of the observed proportions for the two months in each season.

however, *Hypostomus*, like *Ancistrus*, increasingly avoided riffles and remained in pools. At this time, they increased their use of the wood substrates, which offered good cover. *Chaetostomus*, the rarest loricariid in the Rio Frijoles, concentrated in areas where the flow was fast and deep, typically in chutes cut through bedrock formations. In the dry season, however, they also increased their use of deeper areas where flow was slow enough for sand to accumulate. *Rineloricaria* (= *Loricaria*) was small, vertically compressed and much more cryptic than the other three loricariids. This species grazed cobbles and pebbles in riffles where periphyton was relatively abundant. They also rested on sand substrates, and could virtually disappear by burying their outlines. In the dry season, however, even this small cryptic species tended to move into deeper water, as indicated by their increased use of wood and sand substrates.

From the rainy season to the dry season, overlap in substrate use between pairs of the three common loricariid species (*Ancistrus*, *Hypostomus* and *Rineloricaria*) increased or remained the same (Table 1). This result also held when overlap of various size classes was considered (Power in preparation). These findings are congruent with those of Lowe-McConnell (1964, 1967) and Goulding (1980) who found dietary overlap in South American fishes to be higher during the dry season. In contrast, Zaret & Rand (1971) found Panamanian stream fishes to overlap less in diet in the dry season. In the case of *Ancistrus* and *Hypostomus*, increase in overlap probably reflects the importance of cover afforded by submerged wood as water level drops.

Compared with many other tropical streams, the Rio Frijoles is only mildly seasonal (Kramer 1978). Nevertheless, in the dry season, stream substrate under water deeper than 20 cm is reduced to half its rainy season area, while substrate under more than 40 cm is reduced to four tenths (Power 1981). While fishes in other, more seasonal tropical rivers and streams slow or stop their feeding and live off fat stores during the dry season (Lowe-McConnell 1964, 1967, Goulding 1980), loricariids in the Rio Frijoles graze at about the same rate in the dry and

Table 1. Overlap of substrate use by loricariids during dry and rainy seasons.

Loricariid pairs	Proportional similarity	
	Dry season	Rainy season
Ancistrus and *Hypostomus*	0.86	0.63
Ancistrus and *Rineloricaria*	0.71	0.71
Ancistrus and *Chaetostomus*	0.54	0.41
Rineloricaria and *Hypostomus*	0.80	0.79
Rineloricaria and *Chaetostomus*	0.55	0.61
Hypostomus and *Chaetostomus*	0.57	0.72
	Numbers of sightings	
	Dry season	Rainy season
Ancistrus	443	204
Hypostomus	325	128
Rineloricaria	595	253
Chaetostomus	90	33

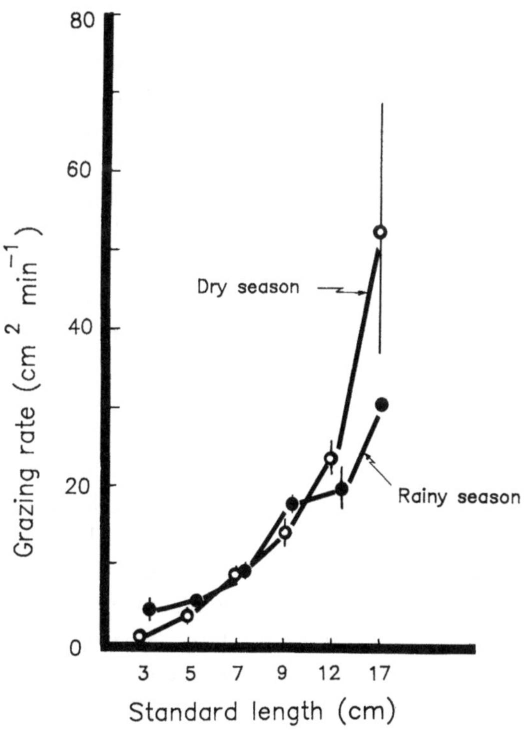

Fig. 3. Grazing rates during the dry season (open circles) and the rainy season (dark circles) for various size classes of *Ancistrus* grazing bedrock substrates in pools. Bars are standard errors. (See Power 1981 for methods.)

the rainy season (Fig. 3). These actively grazing loricariids are food-limited during the dry season, as individuals lose or just maintain weight at this time (Power 1983a). Therefore, their increased overlap in substrate use during the dry season indicates increased interspecific competition for periphyton on the shared substrate. As in their choice of grazing depth, considerations of predator avoidance are more important than foraging considerations in the loricariid responses to seasonal variation in food availability.

Previously grazed patches

Even on initially uniform grazing substrates within a given depth stratum, grazing loricariids must choose among patches in various stages of recovery from previous grazing. In addition to depleting algae, loricariids removed sediment from the substrates they grazed. Many substrates in pools bore layers of sediment, ranging from light dustings to coats several millimeters thick. In the rainy season, this sediment is derived from clay soils that wash into the stream and are deposited during small floods, but are scoured out during large floods (Power 1981). In the dry season, the sediment is derived from decomposing vegetation, which forms a fine, organic-rich (18–24% afdw) rain of outfall onto substrates in quiet water. This sediment, even during the dry season, is not a food for loricariids. *Ancistrus* starved in covered wading pools with dry season sediment lost fat faster than *Ancistrus* starved in clean wading pools (Power 1983b). *Ancistrus* generally avoided grazing areas where sediment was more than a few millimeters deep. Periphyton productivity and standing crops, as indexed by the attached diatoms, were reduced on sedimented substrates (Power 1981). Also, loricariids may have selected cleaner substrates to avoid fouling or abrading their small, entirely ventral gills. Occasionally, however, large *Ancistrus* would open patches in grazing areas where sediment was several millimeters deep. Smaller *Ancistrus* would subsequently graze within these patches, keeping them clear. *Ancistrus* initially opened patches in sediment using a characteristic 'wiggle-thrust' head-down movement that raised a puff of sediment and probably blew much of it away from the fish's respiratory current. In large stream enclosures with relatively high sediment outfall, *Ancistrus* usually began grazing on previously cleared sites (Fig. 4). This selectivity is one example of a response to intermediate-scale spatial variation that seemed largely based on foraging considerations: both tracking higher standing crops of periphyton, and reducing foraging costs.

When they regrazed patches on substrates, loricariids faced various choices: which patches to revisit, how fast to graze within a patch, how closely to crop the periphyton, how frequently to graze rather than perform other activities. These choices may or may not be constrained. For example, changes in overall food availability might cause algivorous fishes to change grazing rates (area covered per time) or severities (biomass removed per area). But if a sharp boundary separated a profitable cropping depth from one at which harvested calories did not balance factors like time lost, wear on teeth or abrasion to gills, fishes might leave fairly constant residues after grazing under a variety of conditions. Microscopic studies, like those of Nicotri (1977) on microalgae grazed by marine gastropods, are necessary to determine the extent to

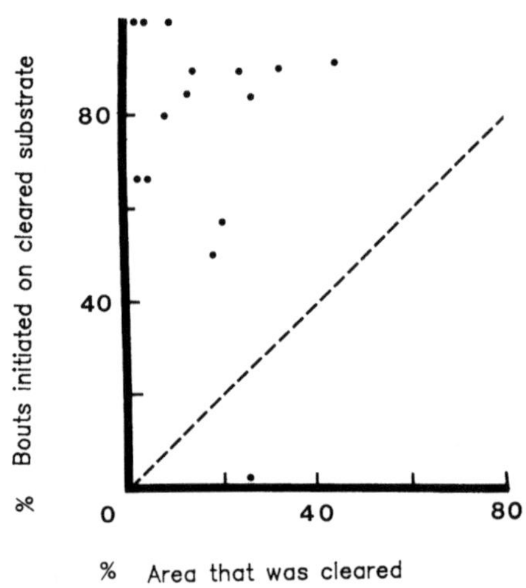

Fig. 4. Selectivity of *Ancistrus* for cleared substrate in enclosures. Points are the percentages of grazing bouts initiated on cleared patches versus the percent of substrate that was clear on a given day in an enclosure. If *Ancistrus* began grazing on sites at random, equal numbers of points should fall above and below the dotted line x = y.

which grazing fishes adjust their cropping severity to different food availabilities.

Grazing severity may be inversely correlated with grazing rate if fish covered area more slowly when they removed periphyton more thoroughly. If these responses are coupled for loricariids, grazing severities may remain fairly constant under a range of conditions. Grazing rates of *Ancistrus* are similar during the dry season, when loricariids are dense, and during the rainy season, when pool densities are reduced by one half or two thirds (Fig. 5). In shaded pools, grazing rates of *Ancistrus* were similar to rates of comparable size classes in sunny pools, where they were on average six times more crowded (although smaller *Ancistrus* graze slightly faster in dark pools). Grazing rates of 6 to 7 cm long *Ancistrus* in stream enclosures where densities were sparse were also similar to their grazing rates in natural stream pool habitats (Fig. 5).

Although grazing rates of *Ancistrus* were similar in different pools, the densities of loricariids varied markedly. The number of grazers sharing a substrate depends both on the number of residents living in a habitat, and on the time individuals allot to foraging on the observed substrate. For territorial grazers, local densities depend on the sizes of the areas grazers defend, and these may change with

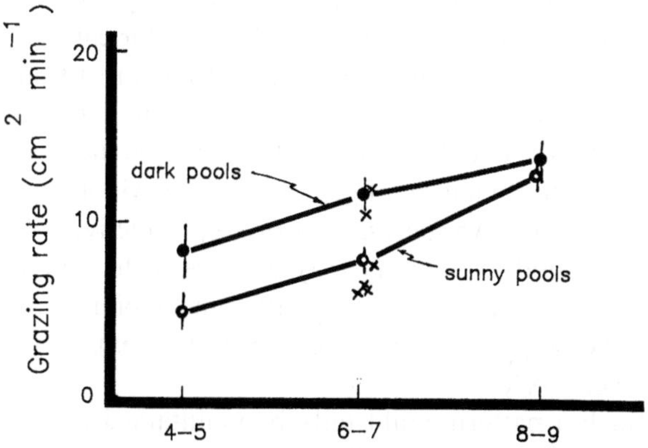

Fig. 5. Grazing rates for *Ancistrus* in two shaded and two sunny pools (dark and open respectively). Bars show 95% confidence limits. Crosses show mean grazing rates for 6–7 cm *Ancistrus* in six stream enclosures built in a sunny reach of the stream and stocked at one sixth the natural density. All grazing rates are pooled over observations made by day and night in the dry and rainy seasons.

fluctuations in food availability. Limpets, *Lottia gigantea*, that graze algal films from rocky intertidal substrates expanded their territories in response to experimental reduction of their food (Stimson 1973). In contrast, food availability did not seem to directly affect territoriality in the ayu, *Plecoglossus altivelis*, a Japanese salmonid that grazes algae from stony substrates in streams. When their densities are sparse (less than three individuals per m^2) certain individuals held territories of about 1 m^2. Those ayu that failed to hold territories did not obtain sufficient food. At higher densities, however, territoriality broke down and most ayu schooled. Surprisingly, growth of schooling individuals was as fast as growth of territory residents in lower densities. Kawanabe (1969) postulated that the territory size of ayu evolved when primary productivity of streams was lower, so that fish are currently defending larger territories than necessary. Alternatively, territory size could be necessary to provide cover from predation or some other advantage for residents.

Nonterritorial grazers at low densities may restrict the area that they graze collectively, as do groups of sheep in pastures. This could prevent them from 'undergrazing', or allowing enough time to elapse between grazing bouts so that food on a site deteriorates. Hunter's (1964) data on voluntary restriction of home range by sheep could be given this interpretation. Similarly, loricariids stocked in large stream enclosures at one sixth their natural densities kept less than half the available substrate grazed and clear of sediment. Possibly, their concentration of grazing effort on this portion of the substrate reduced the detrimental effects of sediment accumulation on attached algae (see Power 1981).

In food-limited populations, however, algae-grazing fishes may return to sites so soon that only scant standing crops of algae have time to develop. The exploitative competition that would lead to 'overgrazing', or returning to sites too soon, might motivate individuals to emigrate from crowded habitats and search for better feeding opportunities elsewhere in their environment. Foraging responses to large-scale variation in periphyton are considered next.

Large scale variation: habitat selection

In the Rio Frijoles, pool habitats for large loricariids alternate with shallow riffles which are partial barriers to larger fishes. Variation in the primary production of attached algae among pools arises because of differences in the density of forest canopy over pools (Fig. 6, bottom). Loricariid density both in terms of individuals and of biomass per area, follows this variation in algal growth rates very closely (Fig. 6, middle). Because higher algal productivity in sunny pools was offset by higher grazer densities, individual loricariids experienced similar food availability in shaded and sunny pools. Therefore, pre-reproductive *Ancistrus* that were marked and periodically recaptured had similar

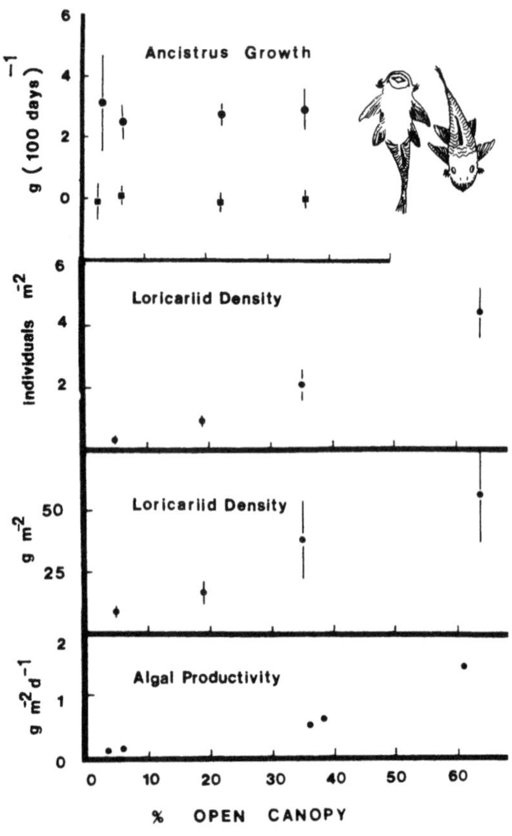

Fig. 6. Periphyton productivity, loricariid densities and somatic growth rates in pools of different canopy covers. Error bars indicate two standard errors. For growth rates, circles indicate averages from the rainy season and squares are averages from the dry season.

somatic growth rates in different pools (Fig. 6, top). In the rainy season, these growth rates (g per 100 days) were (mean $\pm$SE, (N)) 2.7$\pm$0.8 (16), 2.8$\pm$0.3 (63) and 3.1$\pm$0.7 (17) in shaded, half-shaded and sunny pools, respectively. In the dry season, growth rates of pre-reproductive *Ancistrus* were 0.2 $\pm$ 0.2 (23), -0.1 ± 0.2 (24), and 0.0$\pm$0.3 (19) in shaded, half-shaded and sunny pools. Differences among growth rates of pre-reproductive *Ancistrus* in different pools were not statistically distinguishable (Power 1983c).

This pattern, in which food availability was similar in sunny crowded habitats and in shaded, sparsely populated pools, corresponds to the 'ideal free distribution' predicted by Fretwell (1972) and others (Royama 1970) for animals that meet two assumptions. First, if animals have adequate knowledge of the relative qualities of habitats in their environments, and second, if they are free to settle in the best available at any time, then in a relatively empty environment they should settle in the best habitat first, until it is degraded by increased density. When the quality of this site decreases to that of the next best habitat, animals should begin to colonize the second site until it is no better than the third best. As density increases, all available habitats fill in proportion to some balance between their intrinsic quality (reflected by rates of periphyton production for Rio Frijoles loricariids) and degradation of habitat due to competition among residents. Individuals in different habitats should have equal fitnesses, which may be indexed by some combination of growth rates, survivorship, and reproductive success. For pre-reproductive *Ancistrus*, food availability may have been the major factor affecting fitness in different habitats, as survivorship was similar for fish in different pools (Power 1983c).

Such a pattern could only be maintained in the Rio Frijoles if loricariids moved among pools in response to changes in local food availability. The stream was a dynamic environment, and changes were sometimes abrupt, as when trees fell and opened canopies over pools, or when large amounts of sediment were redistributed during floods and pools were created or filled. Loricariids responded quickly to such changes. For example, a new pool created during a large flood was colonized by loricariids within a few weeks of its formation, and within two months it had loricariid densities typical of other pools with its canopy cover (Power 1983c). In addition to abrupt physical changes, more gradual demographic changes such as increased crowding during the dry season occurred. Although most of the 1306 loricariids I marked over a two year period that were resighted were in their home pools, the fish apparently sampled other pools sufficiently to take rapid advantage of new feeding opportunities. By redistributing themselves to track changing food availability among habitats, loricariids damped incipient pool-to-pool variation in periphyton standing crop as it arose.

Conclusions

Fishes grazing periphyton in neotropical streams respond to variation in their food on some scales but not on others. On a small scale, because periphyton is so fine-grained relative to the mouths of grazing fishes, it seems unlikely that they could selectively ingest superior components unless these were clumped, stratified or otherwise spatially segregated. On an intermediate scale, within habitats, avoidance of predators appears to outweigh foraging considerations in the choice of depth strata by pool-dwelling loricariids in the Rio Frijoles. Similarly, large detritivorous cichlids. *Tilapia* (= *Sarotherodon*) *mossambica*, in Lake Sibaya, South Africa also stay in deeper water, where detritus is low in protein. Consequently, these cichlids are nutritionally stunted as adults, although not as juveniles when they feed in shallower areas where detritus is higher in protein (Bowen 1979). Proximity to cover also is an important factor in substrate choice by grazing loricariids, and during the dry season is more important than interspecific competition for food in the choice of grazing substrates by the two most common pool-dwelling species.

On a large scale, however, loricariids track pool to pool variation in periphyton availability very closely. Intrinsic rates of periphyton production in pools are balanced by the collective depletion rates of resident loricariids so that food availability, hence growth rates of pre-reproductive *Ancistrus*, are similar in different pools. Despite demographic, seasonal and occasional nearly catastrophic fluctuations in pool habitats, loricariids remain sufficiently aware of their environment to continually re-evaluate feeding opportunities in their home pools relative to others. By responding quickly, within weeks, to variation that arises on this large scale, these grazers act to respond rapidly so that the long-term availability of food remains fairly constant from pool to pool in an otherwise relatively dynamic environment.

Acknowledgements

I thank Don Kramer, Don Stewart, Bill Dietrich, Tom Zaret, Jim Karr and Paul Angermeier for their helpful comments on various versions of this manuscript. I am also grateful to Bob Paine and Gus van Vliet at the University of Washington for encouragement and useful discussions. This work was supported by a Bacon Fellowship from the Smithsonian Institution and an NSF Doctoral Dissertation Improvement Award (61–7455). I thank the Smithsonian Tropical Research Institute, Panama, for their generous support.

References cited

Alcock, M.B. 1964. The physiological significance of defoliation on the subsequent regrowth of grass-clover mixtures and cereals. pp. 25–41. *In:* D.J. Crisp (ed.) Grazing in Terrestrial and Marine Environments. Blackwell, Oxford.

Blum, J.L. 1956. The ecology of river algae. Bot. Rev. 22: 291–341.

Bowen, S.H. 1979. A nutritional constraint on detritivory in fishes: the stunted populations of *Sarotherodon mossambicus* in Lake Sibaya, South Africa. Ecol. Monogr. 49: 17–31.

Boyd, C.E. 1973. Amino acid composition of freshwater algae. Arch. Hydrobiol. 72: 1–9.

Boyd, C.F. & C.P. Goodyear. 1971. Nutritive quality of food in ecological systems. Arch. Hydrobiol. 69: 256–270.

Calow, P. 1974. Evidence for bacterial feeding in *Planorbis contortus* Linn. Proc. malacol. Soc. Lond. 4: 145–156.

Calow, P. 1975. The feeding strategies of two freshwater gastropods, *Ancylus fluviatilis* Mull and *Planorbis contortus* Linn. (Pulmonata) in terms of ingestion rates and absorption efficiencies. Oecologia 20: 33–49.

Cummins, K.W. 1977. From headwater streams to rivers. American Biology Teacher 39: 305–312.

De Silva, S.S. & E.I.L. Silva. 1979. Biology of young grey mullet, *Mugil cephalus* L., populations in a coastal lagoon in Sri Lanka. J. Fish Biol. 15: 9–20.

Dupree, H.K. & K.E. Sneed. 1966. Response of channel catfish fingerlings to different levels of major nutrients in purified diets. Tech. Paper 9 of the Bureau of Sport Fisheries and Wildlife, U.S. Gov. Printing Office: 1–21.

Dussault, G.V. & D.L. Kramer. 1981. Food and feeding behaviour of the guppy, *Poecilia reticulata* (Pisces: Poeciliidae). Can. J. Zool. 59: 684–701.

Emlen, J.M. 1966. The role of time and energy in food preference. Amer. Natur. 100: 611–617.

Fish, G.R. 1951. Digestion in *Tilapia esculenta*. Nature 167: 900.

Freeland, W.J. & D.H. Janzen. 1974. Strategies in herbivory by mammals: the role of plant secondary compounds. Amer. Natur. 108: 269–289.

Fretwell, S.D. 1972. Populations in a seasonal environment. Princeton University Press, Princeton, 217 pp.

Fryer, G. 1959. The trophic interrelationships and ecology of some littoral communities of L. Nyasa with especial reference to the fishes, and a discussion of the evolution of a group of rock-frequenting Cichlidae. Proc. Zool. Soc. Lond. 132: 153–281.

Fryer, G. & T.D. Iles. 1972. The cichlid fishes of the Great Lakes of Africa. Oliver and Boyd, Edinburgh 641 pp.

Gentile, J.H. 1971. Blue-green and green algal toxins pp. 27–66. *In*: S. Kadis, A. Ciegler & S. Ajl (ed.) Microbial Toxins, vol. 7, Academic Press, New York.

Gerloff, G.C. & F. Skoog. 1954. Cell contents of nitrogen and phosphorus as a measure of their availability for growth of *Microcystis aeruginosa*. Ecology 35: 348–353.

Goulding, M. 1980. The fishes and the forest. Univ. Calif. Press, Berkeley. 280 pp.

Gunn, J.M., S.U. Qadri & D.C. Mortimer. 1977. Filamentous algae as a food source for the brown bullhead, *Ictalurus nebulosus*. J. Fish. Res. Board Can. 34: 396–401.

Hellier, T.R. 1962. Fish production and biomass in relation to photosynthesis in the Laguna Madre of Texas. Pub. Inst. Marine Sci. Univ. Texas 8: 1–22.

Hiatt, R.W. & D.W. Strasburg. 1960. Ecological relationships of the fish fauna on coral reefs of the Marshall Islands. Ecol. Monogr. 30: 65–127.

Hora, S.L. 1933. Respiration in fishes. J. Bombay Natural History Society 36: 538–560.

Hunter, R.F. 1964. Home range behavior in hill sheep. pp. 155–171. *In*: D.J. Crisp (ed.) Grazing in Terrestrial and Marine Environments, Blackwell, Oxford.

Hynes, H.B.N. 1970. The ecology of running waters. Univ. Toronto Press, Toronto. 555 pp.

Kawanabe, H. 1969. The significance of social structure in production of the 'Ayu', *Plecoglossus altivelis* pp. 243–251. *In*: T.G. Northcote (ed.) Symposium on Salmon and Trout in Streams, U.B.C. Fish. Res. Inst., Vancouver.

Keenleyside, M.H.A. 1979. Diversity and adaptation in fish behaviour. Springer-Verlag, New York. 208 pp.

Kramer, D.L. 1978. Reproductive seasonality in the fishes of a tropical stream. Ecology 59: 976–985.

Lassuy, D.R. 1980. Effects of 'farming' behavior by *Eupomacentrus lividus* and *Hemiglyphidodon plagiometopon* on algal community structure. Bull. Mar. Sci. 30: 304–312.

Lowe-McConnell, R.H. 1964. The fishes of the Rupuni savanna district of British Guiana, South America. I. Ecological groupings of the fish species and the effects of the seasonal cycle on fish. J. Linn. Soc. Zool. 45: 103–144.

Lowe-McConnell, R.H. 1967. Some factors affecting fish populations in Amazonian waters. Atas Simp. a Biota Amazonica 7: 177–186.

Lowe-McConnell, R.H. 1975. Fish communities in tropical freshwaters. Longman, London. 337 pp.

MacArthur, R.H. & R. Levins. 1964. Competition, habitat selection, and character displacement in a patchy environment. Proc. Nat. Acad. Sci. USA 51: 1207–1210.

MacArthur, R.H. & E.R. Pianka. 1966. On optimal use of a patchy environment. Amer. Natur. 100: 603–609.

Montgomery, W.L. 1980. The impact of non-selective grazing by the giant blue damselfish, *Microspathodon dorsalis*, on algal communities in the Gulf of California, Mexico. Bull. Mar. Sci. 30: 290–303.

Moriarty, D.J.W. 1973. The physiology of digestion of blue-green algae in the cichlid fish, *Tilapia nilotica*. J. Zool. Lond. 171: 25–39.

Moriarty, D.J.W., J.P.E.C. Darlington, I.G. Dunn, C.M. Moriarty & M.P. Tevlin. 1973. Feeding and grazing in Lake George, Uganda. Proc. L. Soc. Lond. B. 184: 299–319.

Moriarty, D.J.W. & C.M. Moriarty. 1973. The assimilation of carbon from phytoplankton by two herbivorous fishes, *Tilapia nilotica* and *Haplochromis nigripinnis*. J. Zool. Lond. 171: 41–55.

Nicotri, M.E. 1977. Grazing effects of four marine intertidal herbivores on the microflora. Ecology 58: 1020–1032.

Niederholzer, R. & R. Hofer. 1979. The adaptation of digestive enzymes to temperature, season and diet in the roach, *Rutilus rutilus* L. and rudd, *Scardinius erythrophthalmus* L. I. Cellulase. J. Fish Biol. 15: 411–416.

Odum, W.E. 1970. Utilization of the direct grazing and plant detritus food chains by the striped mullet *Mugil cephalus*. pp. 222–240. *In*: J.H. Steele (ed.) Marine Food Chains, Oliver and Boyd, Edinburg.

Ogden, J.C. & P.S. Lobel. 1978. The role of herbivorous fishes and urchins in coral reef communities. Env. Biol. Fish. 3: 49–63.

Payne, A.I. 1978. Gut pH and digestive strategies in estuarine grey mullet (Mugilidae) and tilapia (Cichlidae). J. Fish Biol. 13: 627–630.

Porter, K.G. 1977. The plant-animal interface in freshwater ecosystems. Amer. Sci. 65: 159–170.

Power, M.E. 1981. The grazing ecology of armored catfish (Loricariidae) in a Panamanian stream. Ph.D. Thesis, Univ. Washington, Seattle. 268 pp.

Power, M.E. 1983a. Depth-distributions of armored catfish:

predator-induced resource avoidance? Ecology (in print).
Power, M.E. 1983b. The importance of sediment in the grazing ecology and social interactions of an armored catfish, *Ancistrus spinosus*. Env. Biol. Fish. (in print).
Power, M.E. 1983c. Habitat quality and the distribution of algae-grazing catfish in a Panamanian stream. J. Anim. Ecol. (in print).
Prejs, A. & M. Blaszczyk. 1977. Relationship between food and cellulase activity in freshwater fishes. J. Fish Biol. 11: 447–452.
Roberts, T.R. 1972. Ecology of fishes in the Amazon and Congo basins. Bull. Mus. Comp. Zool. Harv. 143: 117–147.
Roberts, T.R. 1973. Osteology and relationships of the Prochilodontidae, a South American family of characoid fishes. Bull. Mus. Comp. Zool. Harv. 145: 213–235.
Round, F.E. 1964. The ecology of benthic algae. pp. 138–84. *In*: D.F. Jackson (ed.) Algae and Man, Plenum Press, New York.
Round, F.E. 1965. The biology of the algae. Arnold, London. 269 pp.
Royama, T. 1970. Evolutionary significance of predators' response to local differences in prey density: a theoretical study. Proc. Adv. Study Dynamics Numbers Popul. (Oosterbeek): 344–357.
Ruttner, F. 1952. Fundamentals of limnology. Univ. Toronto Press, Toronto. 295 pp.
Schoener, T.W. 1971. Theory of feeding strategies. Ann. Rev. Ecol. Syst. 2: 369–404.
Spoehr, H.A. & H.W. Milner. 1949. The chemical composition of *Chlorella*: effect of environmental conditions. Plant Physiol. 24: 120–149.
Stickney, R.R. & S.E. Shumway. 1974. Occurrence of cellulase activity in the stomachs of fishes. J. Fish Biol. 447–452.
Stimson, J. 1973. The role of the territory in the ecology of the intertidal limpet, *Lottia gigantea* (Gray). Ecology 54: 1020–1030.
Vannote, R.L., G.W. Minshall, K.W. Cummins, J.R. Sedell & C.E. Cushing. 1980. The river continuum concept. Can. J. Fish. Aquat. Sci. 37: 130–137.
Westoby, M. 1974. An analysis of diet selection by large generalist herbivores. Amer. Natur. 108: 290–304.
Wetzel, R.G. 1975. Limnology. Saunders, Philadelphia. 743 pp.
White, T.C.R. 1978. The importance of relative shortage of food in animal ecology. Oecologia 33: 71–86.
Whitton, B.A. 1975. Algae. pp. 81–105. *In*: B.A. Whitton (ed.) River Ecology, Univ. Calif. Press, Berkeley.
Willoughby, L.G. 1977. Freshwater Biology. Pica Press, New York.
Zaret, T.M. & A.S. Rand. 1971. Competition in tropical stream fishes: support for the competetive exclusion principle. Ecology 52: 336–342.

Originally published in Env. Biol. Fish. 9: 103–115

Pterygoplichthys gibbiceps (top specimen) and *Hypostomus carinatus* reduced after a lithograph in Steindachner (1882), Beiträge zur Kenntniss der Flussfische Südamerika's II. Denkschr. Akad. Wiss. Wien 43: 103–146.

Fish communities along environmental gradients in a system of tropical streams

Paul L. Angermeier & James R. Karr
Department of Ecology, Ethology and Evolution, University of Illinois, 606 E. Healey, Champaign, IL 61820, U.S.A.

Keywords: Canopy openness, Community structure, Trophic guilds, Fish distribution, Food availability, Habitat selection, Panama, Seasonality

Synopsis

Fish community structure was examined in 9 forested streams (1–6 m wide) in central Panama during dry seasons over a 3 year period. Study regions varied in annual rainfall, degree of canopy shading, and topographical relief. Benthic invertebrates were more abundant in riffles than in pools and more abundant in early (January) than late (March) dry season. In addition, benthos abundances were negatively correlated with canopy shading among study regions. Terrestrial invertebrate abundances were greater in January than March and were correlated with stream width. Fishes were assigned to 7 feeding guilds (algivores, aquatic insectivores, general insectivores, piscivores, scale-eaters, terrestrial herbivores, omnivores) on the basis of similarity of gut contents. Four species exhibited marked dietary shifts with increasing size. Distributions of feeding guilds (biomass) among habitats and streams were not generally correlated with availabilities of their major food resources. All feeding guilds except aquatic insectivores were most concentrated (biomass per area) into deep pools. Densities of algivores and terrestrial herbivores increased with stream size, but the density of aquatic insectivores declined. Species richness of feeding guilds increased with stream size and canopy openness. The proportion of fish biomass supported by algae and terrestrial plant material increased with stream size, while that supported by aquatic and terrestrial invertebrates declined. Small fishes (< 40 mm TL) were most abundant in pools of small streams. Terrestrial predators appeared to be more important than food availability in determining distributions of fish among habitats. However, trophic diversity of fish communities may be related to tne reliability of available food resources.

Introduction

A central goal of community ecology is to understand mechanisms and processes responsible for differences and similarities among communities. One approach is to compare communities occurring along environmental gradients such as physiological stress or resource availability. This approach provides valuable basic data and encourages the development of testable hypotheses. Such hypotheses will, in turn, lead to more definitive studies, including experimental manipulations.

Existing models of the structure and dynamics of stream communities are largely based on patterns observed in forested temperate streams, for example, the stream continuum hypothesis (most recently presented in Vannote et al. 1980). Briefly, this hypothesis holds that relative abundances of various food types vary predictably with stream size; relative abundances of consumer guilds are correlated with those of their major food resources. The model emphasizes the importance of trophic

Thomas M. Zaret (ed.), Evolutionary ecology of neotropical freshwater fishes. ISBN 90 6193 823 6

function and food availability to the distribution and abundance of stream organisms. Food availability and consumption patterns are regulated by fluvial geomorphic processing of organic materials in the stream. The stream continuum hypothesis, however, is less sensitive to variables such as water chemistry and the complex biotic interactions typical of streams. A conceptual model of stream communities should incorporate these other factors (Karr & Dudley 1981), especially for the assessment of human impacts on streams (Karr 1981).

Because virtually all stream models have been developed from a limited data base and geographic perspective, tests are needed in other regions. Our main objectives in this study, therefore, were to identify patterns of food and habitat availability and predator abundance, and relate those patterns to the distribution and abundance of fishes in central Panama streams. Our hope was to identify basic patterns and to generate hypotheses for future study that will elucidate the organizational processes of streams and their biotic communities.

Materials and methods

Study sites

Central Panama is a narrow isthmus extending 70 km from the relatively dry Pacific Coast to the wetter Atlantic Coast. Rainfall is seasonal with a dry season beginning in late December or early January and extending to early April. Monthly and yearly rainfall totals vary among years (Table 1), but the January through March period is always the driest. The nearest rainfall stations (Fig. 1) to our study sites are at Gamboa with an average annual rainfall of 2200 mm and Frijoles with 2680 mm.

We studied nine streams along the Pipeline Road in Parque Nacional Soberania in the Rio Chagres drainage of central Panama. This tract of lowland rainforest has remained largely undisturbed since the early 1900's. The study streams generally drain mature second-growth forest except for a few areas of immature forest as noted below.

Three groups (regions) of streams were identified on the basis of rainfall, watershed topography, and forest canopy characteristics, while streams within regions varied in size (Table 2). Region 1 contained only one stream, Quebrada Juan Grande, which drained recently disturbed forest on land with rolling topography. Region 2 contained the Rio Frijoles and 3 tributary streams: Rio Frijolito, Tower Creek, and Rio Limbo. Region 2, about 3 km from Region 1, was characterized by rolling topography and structurally mature rainforest except for a small area of immature forest upstream from sample sites on the Rio Frijoles. Region 3 contained Rio Mendosa and 2 tributaries, Rio Sirystes and Tayra Creek, as well as Rio Pilon. All 4 streams drained structurally mature forest in rugged topography. The Mendosa watershed and the Rio Pilon were about 7 and 10 km, respectively, from Region 2. Exact data are not available but rainfall at nearby sites indicate that Region 3 receives approximately 500 mm more rainfall annually than Region 1, with Region 2 intermediate between Regions 1 and 3. Most of the difference in

Table 1. Rainfall (mm) at nearby Barro Colorado Island during study periods with annual totals and variation (1979–81 data obtained courtesy of D. Windsor, Smithsonian Tropical Research Institute; 1941–1980 data courtesy of Panama Canal Meteorological Bureau).

Year	Dry season				Year total
	January	February	March	Total	
1979	5	30	8	43	2654
1980	112	48	5	165	2096
1981	399	20	61	480	4633
$\bar{x} \pm SD$ (1941–1980)	63 ± 67	32 ± 34	26 ± 27	122 ± 96	2563 ± 382

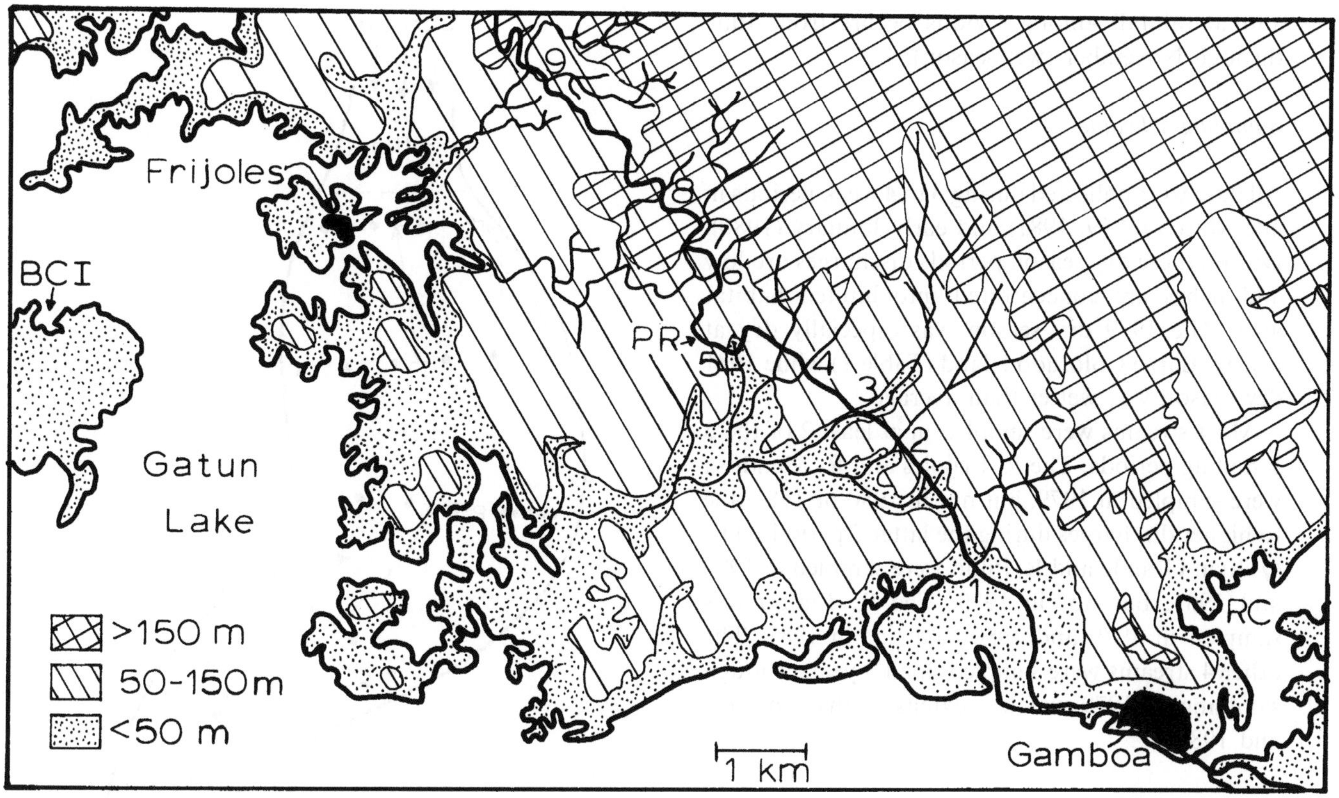

Fig. 1. Map of Pipeline Road (PR) area in central Panama showing Barro Colorado Island (BCI), Rio Charges (RC) and Gatun Lake. Study sites are located on Quebrada Juan Grande (1), Rio Frijolito (2), Rio Frijoles (3), Rio Limbo (4), Tower Creek (5), Rio Mendosa (6), Rio Sirystes (7), Tayra Creek (8), and Rio Pilon (9).

Table 2. Average stream width and percentage ($\bar{x} \pm$ SD) of sample points with selected habitat features in 9 Panama streams. Percentages were calculated over all available samples except in R. Limbo, an intermittent stream.

Stream	Ave. width (m)	Depth (cm)			Perceptible current	Substrate				Leaf cover	
		1–29	30–59	>60		sand	gravel	pebble	bedrock	March	January
Region 1											
Juan Grande	2.0	87 ± 2	10 ± 1	3 ± 3	21 ± 13	3 ± 0	75 ± 2	5 ± 4	0 ± 0	19	4
Region 2											
Frijoles	5.1	70	22	8	7	14	44	17	21	16	—
Frijolito	4.3	79 ± 3	15 ± 1	6 ± 2	24 ± 12	14 ± 1	45 ± 11	30 ± 5	7 ± 1	22 ± 3	6
Limbo*	2.3	77	21	2	9	9	60	12	0	—	10
Tower	1.3	89 ± 7	8 ± 3	3 ± 4	6 ± 6	16 ± 13	48 ± 18	2 ± 3	0 ± 1	50 ± 24	12
Region 3											
Mendosa	5.6	73	16	11	19	21	32	20	10	19	—
Sirystes	3.7	78 ± 5	17 ± 4	5 ± 1	21 ± 5	8 ± 2	33 ± 6	25 ± 1	20 ± 5	24 ± 8	15
Pilon	2.7	87 ± 3	13 ± 3	0 ± 0	29 ± 6	3 ± 0	63 ± 16	18 ± 8	18 ± 4	27	11
Tayra	1.6	92 ± 1	7 ± 1	0 ± 0	16 ± 10	7 ± 6	41 ± 12	25 ± 4	15 ± 5	34 ± 25	10

* Data from 1980 only.

annual rainfall among regions occurs from June through November (wet season).

Sampling schedule

All data were collected during 3 to 4 week periods in late March (1979, 1981) and early January (1980). During these dry season periods discharges are low and fishes are concentrated into limited habitat area. The 1979 dry season was especially dry at nearby Barro Colorado Island (Table 1) while 1981 dry season was wetter than average.

Most streams were sampled in at least 2 of the 3 sampling periods. The R. Frijoles and R. Mendosa were sampled in 1979, but not in 1980 or 1981 due to higher discharges during the latter 2 periods. The Q. Juan Grande and R. Pilon were sampled in 1980 and 1981, but not in 1979. The R. Limbo was sampled in 1979 and 1980, but not in 1981. The other 4 streams were sampled in all 3 years, though sampling in 1979 was less extensive than in 1980 and 1981.

Sampling methods

Habitat structure. – One to three days prior to fish sampling, structural features of each stream channel were systematically assessed along a series of transects. The method used (Fig. 2) was modified from that used by Gorman & Karr (1978). Distances between transects (T) ranged from 1.0 to 3.3 m while distances between habitat points along transects (P) ranged from 0.5 to 1.0 m. Larger intervals between transects and points were used in large streams than in small streams. At each habitat point, depth, current, bottom characteristics, and cover availability were evaluated. Depth was measured to the nearest centimeter. Surface current was gauged by the distance that the water was driven up the edge of a vertically held meter stick. When that distance exceeded 1.5 mm, current was considered perceptible. Mineral substrates were classified as silt, sand, gravel, pebble, rock, bedrock, or hard clay largely in accord with the classification given in Hynes (1970). Organic substrates included leaves and other organic litter. Cover features, including overhanging vegetation, undercut banks, branch, log, and rock structures, were noted. From 76 to 228 m of each stream channel were evaluated by this method, depending on stream size and sample period; the corresponding number of habitat points evaluated from each stream ranged from 84 to 389. Longer reaches were sampled in larger streams, and sampling in 1980 and 1981 was more extensive than in 1979. Area of a given stream reach was estimated using the number of habitat points sampled and the distances between points and transects.

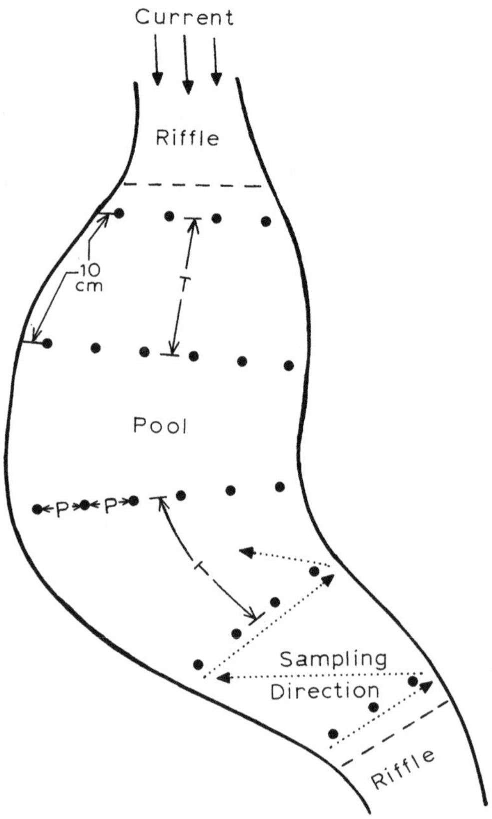

Fig. 2. Diagrammatic representation of the technique used to assess habitat structure of streams. P in the diagram varied from 0.5 to 1.0 m, while T varied from 1.0 to 3.3 m depending on stream size. At each point (darkened circles) several habitat variables were evaluated (see text). The first point of each transect was located 10 cm from the water's edge. This technique was repeated in each habitat unit (i.e. pool, riffle, raceway) of all study reaches. Dashed lines in the figure represent approximate delineations between habitat units.

Canopy closure. – Canopy closure was measured in 1980 and 1981 over streams sampled during those periods. At the center of each habitat transect the percent canopy closure was visually estimated and assigned to one of 5 categories (0–20%, 20–40%, etc.). The height at which the canopy first reached maximum closure was also estimated to the nearest 1.5 m.

Aquatic invertebrates. – Benthic invertebrates were sampled during the day from representative substrates in midstream using a Surber sampler (0.093 m^2) with 240 μm mesh. Two riffles and 2 shallow pools were generally sampled from each stream. Fifteen such samples were collected in March 1979, while 30 were collected in January 1980. Sites of invertebrate samples were nearby and similar to those used for fish sampling. All invertebrates were preserved immediately in 15% formalin. Samples were subsampled using a device similar to that of Waters (1973), then sorted to order and counted using a dissecting microscope.

Terrestrial invertebrates. – Several types of wide-mouth containers (e.g. dishpans, inflatable wading pools) were used to catch invertebrates that fell or flew onto the water surface. Containers were placed in the stream channel (though not usually in the water) and partially filled with a solution of dish-washing detergent (1–2%). Canopy characteristics above the containers were recorded. Containers were left for one to three 24 h periods, after which the contents were strained through a 240 μm mesh net, and preserved in 10% formalin. One to 4 such samples were collected from each stream; container openings ranged from 0.088 to 0.374 m^2 in area. Invertebrates were later sorted and counted under a dissecting microscope.

Fishes. – Stream reaches were divided into habitat units (e.g. pool, riffle, pool, etc.) for fish sampling. Before sampling, block nets were placed at natural barriers in the channel (Fig. 2) on the up- and downstream ends of a habitat unit. Non-stationary debris (e.g. branches, logs, etc.) and rocks that protruded into the water column were removed from the channel to facilitate seining. Four to eight habitat units were sampled from each stream studied in 1979, while 6–16 units were sampled from each stream studied in 1980 and 1981. Overall, 162 samples of habitat units were collected. An effort was made to acquire replicate samples of the predominant habitats (i.e. deep pool, raceway, riffle) in each stream. More habitat units were sampled in large streams than in small streams.

Fish sampling was performed during the day (0800–1800 h) using a 1.4 × 6.7 m bag seine with 0.48 cm mesh. The seine was drawn through each pool and emptied 3 times, then fish were counted. Riffles were sampled by thoroughly disturbing the substrates down to 10 cm by kicking while moving toward a stationary net. The stationary net was relocated at 2–3 m intervals through a given riffle, and each riffle was sampled 3 times. This 'kicking' procedure was more effective at capturing riffle fishes than the typical seining procedure used in pool sampling. Methods used here underestimate relative abundances of secretive fishes such as *Synbranchus*.

Fish were generally identified to species using keys prepared by J.D. McPhail. For *Astyanax ruberrimus* and *Bryconamericus emperador* data were pooled for analysis, but a subsequent examination of preserved collections indicated that *Astyanax* was more common than *Bryconamericus*. Most fish were measured to the nearest millimeter, total length (TL), then released. Fish to be used for stomach analyses were anesthetized in tricaine-methanesulfonate to prevent regurgitation, then preserved in 20–25% formalin. Weights of released fish were estimated from length-weight regressions generated from preserved specimens. These weight estimates are not comparable to live weights since fish specimens shrink during preservation (Parker 1963) but estimates from preserved specimens do permit comparisons of fish weights for collections that are preserved and stored similarly.

Fish food habits. – The 3 fish collections contained enough individuals of 26 species for analysis of food habits. An exception was *Synbranchus marmoratus*. Contents of foreguts and stomachs were assigned to 6 categories: algae, aquatic invertebrates, terrestrial invertebrates, terrestrial plant material, fish, and fish scales. Percent volume comprised by each food type was estimated to the nearest 5% for each gut. Averages (over individuals) of these percentages were used to assign fishes to feeding guilds. An effort was made to include individuals of various sizes and stream origins in assessing food habits. Weighted pair-group cluster analysis with arithmetic averages (Sokal & Sneath 1963) using Horn's (1966) ecological-overlap index was used to group species and size classes into feeding guilds. Average diet proportions for each size class or species were used to calculate the overlap values.

Results

Physical features

Habitat structure. – All streams were characterized by well defined pool-riffle sequences, with pool habitats comprising most of the channel area. Current through pools and raceways was usually not perceptible (Table 2) during this study, but wet season currents are substantial in all habitats. Although gravel was the most common substrate in all streams, the frequency of rocks and bedrock increased from Region 1 through Region 3 (Table 2). The proportion of the stream bottom covered with leaves decreased with increasing stream size in Regions 2 and 3. The stream in Region 1 (immature forest) consistently contained fewer leaves than streams of similar size in Regions 2 (Tower, Limbo) and 3 (Pilon, Tayra). Leaf coverage was consistently higher in March than in January.

Stream reaches were divided into habitat units (i.e. pool, raceway, riffle) on the basis of depth and current characteristics. Riffles were shallow (less than 12 cm mean depth) with turbulent flow. Pools and raceways had negligible current; maximum depth of pools was at least 50 cm, while raceways were shallower with mean depth less than 17 cm.

Canopy coverage. – The proportion of the stream channel open to direct sunlight increased with stream size (Table 3). Region 1 forest canopy was more open than that over Regions 2 and 3. Furthermore, canopy height increased from Regions 1 through 3. The difference in canopy height between Region 1 (immature forest) and Region 2 (mature forest) seemed to be largely due to differences in forest age, while the difference in canopy height between Regions 2 and 3 (both mature forest) seemed to be largely due to topographical differences. Streams in Region 3 were more completely shaded than those in Region 2 due to the higher canopy associated with the steeper banks in Region 3.

Table 3. Stream widths and percentage ($\bar{x} \pm SD$) of canopy estimates assigned to categories of canopy closure and height over 9 Panama streams. Percentages were computed from January 1980 and March 1981 data except for R. Limbo, which was only sampled in 1980.

Stream	Ave. width (m)	<40% Closed	Max. closure below 6 m
Region 1			
Juan Grande	2.0	17 ± 8	51 ± 23
Region 2			
Frijolito	4.3	10 ± 5	23 ± 14
Limbo	2.3	12	15
Tower	1.3	1 ± 2	7 ± 1
Region 3			
Sirystes	3.7	10 ± 6	0 ± 0
Pilon	2.7	3 ± 0	11 ± 7
Tayra	1.6	0 ± 0	4 ± 6

Food availability

Benthic invertebrates. – The most abundant invertebrates were Ephemeroptera, Coleoptera and Diptera larvae (Table 4). Taxonomic composition (at ordinal level) of the benthos did not differ markedly among streams. Total benthos abundances did vary significantly through space and time. Three variables accounted for 70% of the variance in benthos densities (Fig. 3) in a multiple regression analysis. This analysis estimates the effects of individual independent variables while holding others constant statistically. Invertebrates were about 3.5 times more abundant in riffles than pools, and about 3 times more abundant in Region 1 than Region 3. Benthos were also more abundant in January 1980 than March 1979.

Terrestrial invertebrates. – Invertebrates falling into collecting pans were mostly Diptera and Coleoptera adults. Two variables in a multiple regression ana-

Table 4. Percentages of total benthic invertebrate numbers comprised by 3 predominant taxa in 3 streams. Data are from Surber samples collected in January 1980.

Stream (n)	Ephemeroptera		Coleoptera		Diptera	
	$\bar{x}$	SD	$\bar{x}$	SD	$\bar{x}$	SD
Queb. Juan Grande (4)	52.6	13.3	19.7	6.3	18.7	11.9
R. Frijolito (6)	35.5	7.1	22.7	6.8	27.3	8.6
R. Sirystes (4)	35.8	10.0	19.7	8.2	24.7	12.6

lysis accounted for 65% of the variance in the abundance of terrestrial invertebrates (Fig. 4). Abundances increased significantly with stream size and were about 6 times greater in January 1980 than March 1981.

Fish community

Twenty-seven fish species were captured during the study, with little variation among sampling periods in species' occurrences in each stream (Table 5). Several species known to occur in streams of the area were not captured (M. Power, personal communication). Although probably common in most streams (J. Graham, personal communication), only 3 specimens of *Synbranchus marmoratus* were collected. This species was excluded from further analysis since it was not adequately sampled with our methods.

No species were found exclusively in Region 3, but 8 species found in Region 2 were absent from Region 3 (Table 5). Species numbers increased with

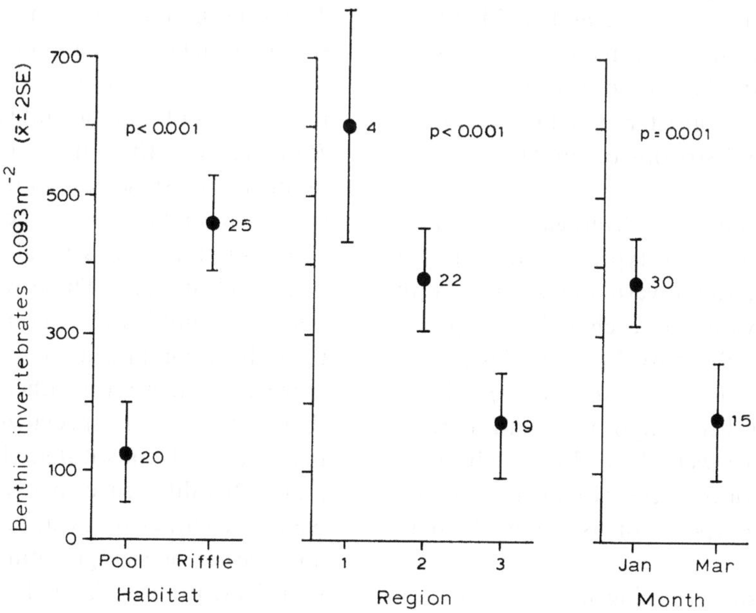

Fig. 3. Number of benthic invertebrates in Surber samples from streams along the Pipeline Road. Numbers beside means indicate the sample sizes associated with values taken by independent variables (horizontal axes). Means and standard errors are estimated from a multiple regression model.

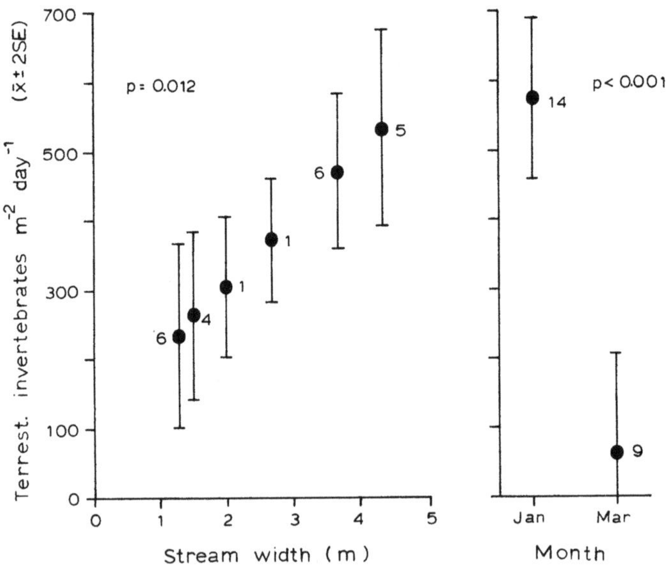

Fig. 4. Number of terrestrial invertebrates falling into collecting devices in streams along the Pipeline Road. Numbers beside means indicate sample sizes associated with values taken by independent variables. Means and standard errors are estimated from a multiple regression model.

stream size in both Regions 2 and 3. Numbers of species inhabiting the R. Frijoles and R. Mendosa were probably underestimated due to the relatively small amount of sampling done in those streams. Streams of Region 2 consistently supported more species than Region 3 streams of similar size.

Feeding guild assignment. – Average % volumes comprised by major food types (algae, aquatic invertebrates, terrestrial invertebrates, etc.) were computed over individuals of each of 30 fish species and size groupings (Table 6). Four species (*Pimelodella, Aequidens, Brycon, Astyanax*) exhibited substantial shifts in diet composition with increasing size. These species were divided on the basis of total length into 2 (or 3, in the case of *Brycon*) size groupings for the purposes of assigning them to feeding guilds.

Each species or size grouping was assigned to a feeding guild largely on the basis of a cluster analysis using Horn (1966) overlap measures (Fig. 5). Algivore, aquatic insectivore, and general insectivore guilds were composed of species or size groupings overlapping in food use by at least 90%. Algivores and aquatic insectivores specialized in eating attached algae and aquatic invertebrates, respectively. General insectivores consumed similar proportions of aquatic and terrestrial invertebrates. *Roeboides* was the only species that ate substantial numbers of fish scales, while *Hoplias* was the only species specializing on other fish. Five species or size groupings ate several food types (including plant and animal). These were all assigned to the omnivore guild, even though the proportions of their diets comprised by various food types were more variable than in other guilds. *Brycon* greater than 130 mm TL specialized on terrestrial plant material, and constituted the terrestrial herbivore guild. Diet differences between March and January were not sufficient to affect guild assignments of any species or size grouping. Nearly all guts contained identifiable food. In only 2 genera (*Rivulus* and *Hoplias*) were guts empty in more than 10% of individuals (Table 6).

Table 5. List of fish species used in food habits and fish distribution analyses, and total number of species per stream. Species entries indicate number of sampling periods species was captured. Numbers in parentheses indicate number of sampling periods that stream was sampled. Total species numbers were computed over the 3 sampling periods.

	Regions 1	2				3			
	Juan Grande (2)	Frijoles (1)	Frijolito (3)	Limbo (2)	Tower (3)	Mendosa (1)	Sirystes (3)	Pilon (2)	Tayra (3)
Characidae									
Cheirodon gorgonae	2		1	2					
Gephyrocarax atricaudata	2	1	3	2	3			2	
Astyanax ruberrimus	2	1	3	2	3	1	3	2	
Bryconamericus emperador	2	1	3	2	3	1	3	2	
Hyphessobrycon panamensis	2	1	3	2	3				
Roeboides guatemalensis	2	1	2	1				2	
Brycon petrosus		1	3			1	3	2	
Lebiasinidae									
Piabucina panamensis	2	1	3	2	3	1	1	1	3
Erythrinidae									
Hoplias microlepis	2	1	2	1	2				
Hypopomidae									
Hypopomus occidentalis	2	1	3	1	2		3	2	3
Pimelodidae									
Rhamdia wagneri	2		3	1					
Pimelodella chagresi		1	3			1	2	2	
Imparales panamensis			3		1				
Trichomycteridae									
Trichomycterus striatum	1	1	2				3	2	3
Loricariidae									
Hypostomus plecostomus		1	3	1					
Chaetostomus fischeri		1	2			1	2	2	
Ancistrus chagresi	1	1	3	1					
Rhineloricaria uracantha	2	1	3		1	1	2		
Cyprinodontidae									
Rivulus brunneus	2		3	2	3	1	3	2	3
Poeciliidae									
Poecilia sphenops	2	1	3	2	2	1	3	1	1
Neoheterandria tridentiger	2	1	3	2	3		3		
Brachyraphis episcopali	2		2	2	2	1	3	2	3
Brachyraphis cascajalensis	2	1	3	2	3	1	3		
Cichlidae									
Aequidens caeruleopunctatus	2	1	3	2	3	1	2	2	
Geophagus crassilabris		1	3		3				
Cichlasoma panamensis	2								
Total species	20	20	25	18	16	12	15	14	6

Guild distributions. – Multiple regression analyses were used to test for correlations between estimates of guild biomass densities (g m^{-2}) and 4 independent environment variables: month, habitat, region and stream width. No significant differences in feeding guild biomass were observed between January and March samples. Habitat type was the variable most often correlated with guild distribution. Six of the 7 feeding guilds were most concentrated in pools and least concentrated in riffles (Fig. 6). Similar densities of aquatic insectivores were found in riffles, raceways, and pools. Regional effects on feeding guild density were observed for *Roeboides* (scale-eater) and *Hoplias* (piscivore). *Roeboides* abundance increased from Region 3 to Region 1 ($p = 0.001$) while *Hoplias* was never captured in Region 3 (Table 5). Species richnesses of the algivore, aquatic insectivore, and general insectivore guilds were consistently higher for Region 2 than Region 3 streams of similar size (Fig. 7). Apparent declines in species richness for these guilds in the largest streams (Frijoles and Mendosa) are probably not real, but artifacts of less intensive sampling relative to other streams.

Effects of stream size were observed for both biomass density and species richness patterns of feeding guilds. Both density (Fig. 8) and species richness (Fig. 7) of algivores increased with increasing stream size. Aquatic insectivore density declined downstream (Fig. 8) even though species richness increased (Fig. 7). Terrestrial herbivores (large *Brycon*) were more abundant in larger streams (Fig. 8), and, indeed, never occurred in streams without deep pools. The number of types of omnivores present also increased with stream size (Fig. 7), while the number of general insectivore species remained relatively invariant across stream size.

In summary, distributions of guild biomass among habitats and streams were complex. Algivores reached greatest densities in large pools of Region 2, but their presence was sporadic and variable. Aquatic insectivores became less abundant in larger streams, but were found regularly in all habitats. More species of aquatic insectivores occurred in Region 2 than Region 3. Scale-eaters (*Roeboides*) achieved highest densities in deep pools

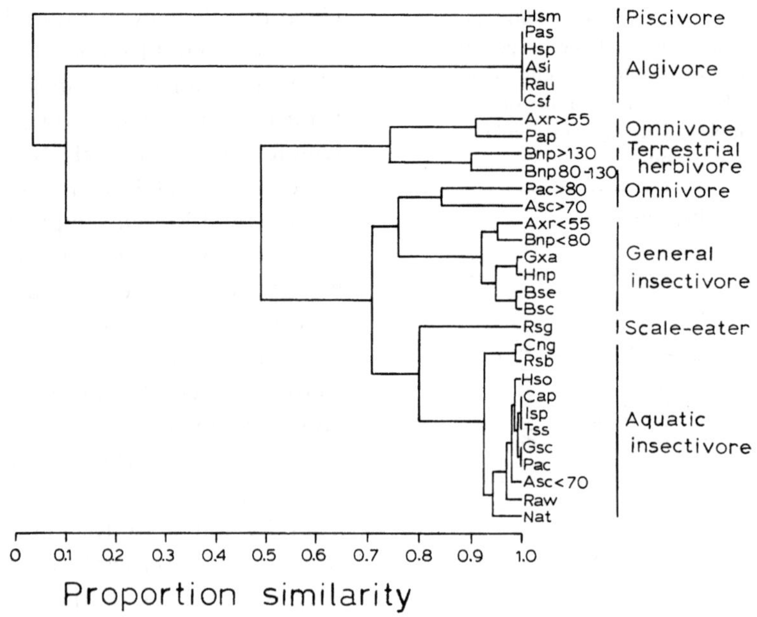

Fig. 5. Dendrogram used to assign fish species and size groupings to feeding guilds. Clusters are based on Horn (1966) overlap values of diet compositions as listed in Table 6.

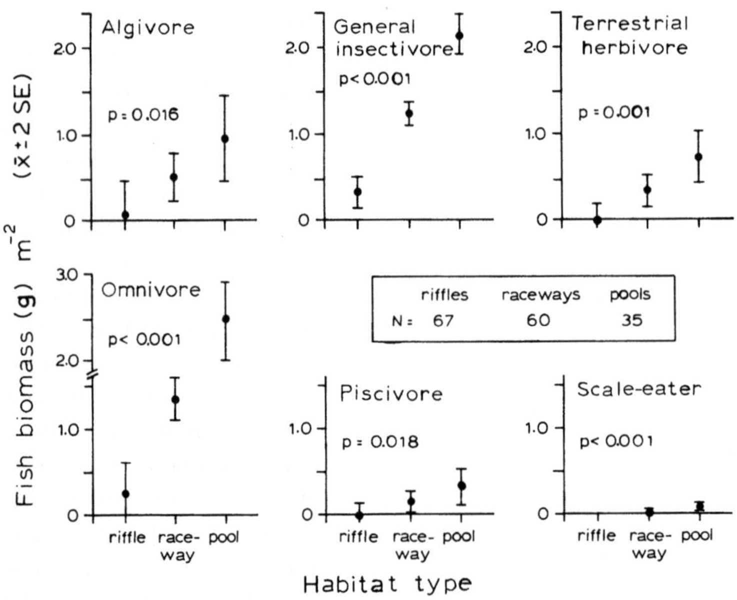

Fig. 6. Distribution of fish feeding guilds among habitat types in streams along the Pipeline Road. Means and standard errors are estimated from multiple regression models.

Table 6. Percent volumes ($\bar{x} \pm$ SE) of various food types found in guts of fish species from Pipeline Road streams for March 1979 and January 1980. Acronyms beside species names are used as reference in Figure 5. Four genera (*Pimelodella*, *Aequidens*, *Astyanax* and *Brycon*) are listed in separate size classes.

Guild/genus (code)	% Algae	% Aquatic invertebr.	% Terrest. invertebr.	% Terrest. plant mat.	% Fish	% Fish scales	No. of guts with food
Algivores							
Ancistrus (Asi)	100 ± 0						5
Hypostomus (Hsp)	100 ± 0						3
Chaetostomus (Csf)	100 ± 0						3
Rhineloricaria (Rau)	100 ± 0						5
Poecilia (Pas)	100 ± 0						5
Aquatic insectivores							
Hypopomus (Hso)	3 ± 2	97 ± 2		1 ± 0.3			37
Trichomycterus (Tss)		90 ± 7		0.3 ± 3			18
Pimelodella (Pac)							
< 80 mm		99 ± 1	1 ± 1				9
Rhamdia (Raw)		94 ± 5	1 ± 1	3 ± 3		2 ± 2	13
Imparales (Isp)		100 ± 0					12
Aequidens (Asc)							
< 70 mm	0.5 ± 0.4	95 ± 2	2 ± 1	0.2 ± 0.2	2 ± 2		30
Geophagus (Gsc)		98 ± 1	1 ± 1	1 ± 1			12
Cichlasoma (Cap)		90 ± 10					10
Rivulus (Rsb)		87 ± 7	15 ± 6	1 ± 0.4	2 ± 2		26
Neoheterandria (Nat)	9 ± 5	88 ± 6	3 ± 3				30
Cheirodon (Cng)	0.8 ± 0.8	87 ± 13	13 ± 13				6
General insectivores							
Brachyraphis							
cascajalensis (Bsc)	2 ± 2	47 ± 7	51 ± 6	0.2 ± 0.2			24
B. episcopali (Bse)		49 ± 6	49 ± 6	1 ± 1			39
Hyphessobrycon (Hnp)	1 ± 1	32 ± 6	61 ± 7	6 ± 3			35
Gephyrocarax (Gxa)		25 ± 6	73 ± 6	3 ± 1			38
Brycon (Bnp)							
< 80 mm		54 ± 8	31 ± 7	15 ± 6			26
Astyanax (Axr)							
< 55 mm	7 ± 4	45 ± 7	39 ± 6	9 ± 3			33
Omnivores							
Aequidens (Asc)							
≥ 70 mm	1 ± 1	45 ± 16	6 ± 4	22 ± 11	25 ± 16		7
Pimelodella (Pac)							
≥ 80 mm		46 ± 18	16 ± 15	38 ± 15			4
Piabucina (Pap)	19 ± 5	14 ± 5	39 ± 7	20 ± 5	6 ± 3	0.2 ± 0.2	41
Brycon (Bnp)							
80–130 mm		10 ± 5	18 ± 6	72 ± 7			28
Astyanax (Axr)							
≥ 55 mm	10 ± 4	6 ± 3	39 ± 6	47 ± 6		0.3 ± 0.2	38
Terrestrial herbivores							
Brycon (Bnp)							
> 130 mm		0.4 ± 0.4	6 ± 3	89 ± 4	4 ± 3		14
Piscivores							
Hoplias (Hsm)					100 ± 0		6
Scale-eaters							
Roeboides (Rsg)		58 ± 9	3 ± 3	4 ± 4		27 ± 9	17

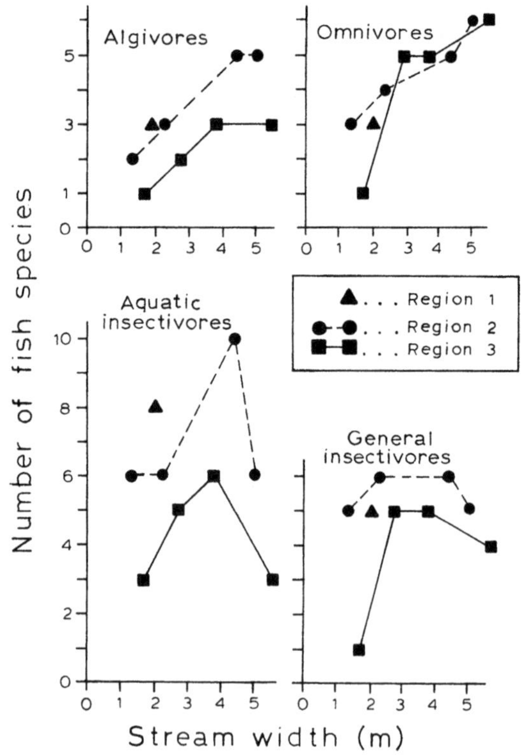

Fig. 7. Species richness patterns of fish feeding guilds illustrating effects of region and stream size. Species numbers were computed over all samples. Guild assignments are given in Table 6.

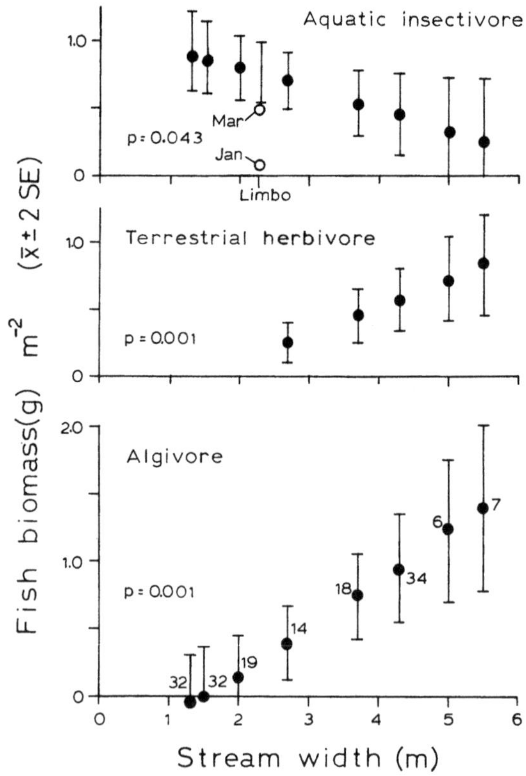

Fig. 8. Effects of stream size on biomass densities of fish feeding guilds. Numbers beside means in the bottom graph indicate the number of habitat units sampled in each stream. Means and standard errors are estimated from multiple regression models. Open circles in the top graph indicate biomass values observed in R. Limbo in March 1979 and January 1980.

of Region 1. Piscivores (*Hoplias*) were most common in pools of Region 1 and 2. General insectivores were abundant in most pools, particularly Region 2. Terrestrial herbivores (large *Brycon*) occurred sporadically, but only in deep pools. Omnivores were also widespread and abundant especially in pools.

Fish distributions. – Fish distributions were also affected by severe seasonality in discharge. All stream discharges were low during March 1979, but only R. Limbo (Region 2) was reduced to widely separated pools. Fish density was comparable to that expected on the basis of stream size, with most fish biomass being small general insectivores. In January 1980, R. Limbo's discharge was substantial, but fish density was low compared to other streams. This effect was observed for aquatic insectivores (Fig. 8) as well as general insectivores and omnivores.

Relative importance of food types. – The importance of 5 major food resources (algae, aquatic invertebrates, fish, terrestrial invertebrates, terrestrial plants) was assessed by attributing fish biomass to a given food resource in proportion to the volume occupied by that food resource (from Table 6) in fishes' guts. For example, we assumed that 2%, 97%, and 1% of the biomass of *Hypopomus* was supported by algae, aquatic invertebrates, and terrestrial plant material respectively, while 100% of the biomass of *Imparales* was supported by aquatic invertebrates. Allocation of fish biomass to various food resources was performed for each stream each

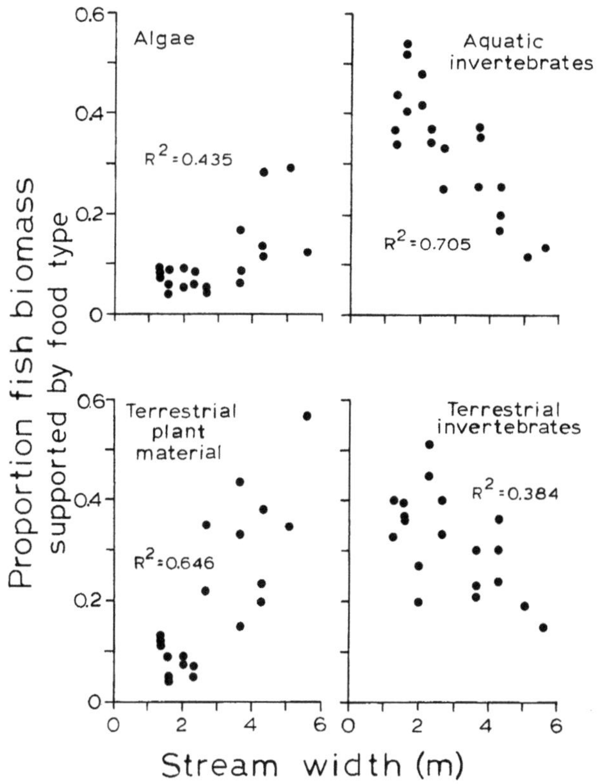

Fig. 9. Effect of stream size on the relative importance of 4 major food resources in supporting fish biomass. Each point represents data pooled from all habitats sampled from a stream during a sample period. All 4 regressions are significant at $p < 0.004$.

time it was sampled. These data were then converted to proportions to yield relative importance values of each food resource in supporting the fish community.

Relative importance values of each food resource were regressed on 3 independent environmental variables: month, region and stream width. No food resource differed in the proportion of the fish biomass it supported between January and March samples. A regional effect was significant only for the relative importance of fish (including scales) as food. Proportion of fish biomass supported by other fish increased from Region 3 through Region 1 ($p < 0.001$; $R^2 = 0.628$). Stream width was the best predictor of the importance of food resources to the fish community (Fig. 9). Both terrestrial and aquatic invertebrates declined in relative importance as stream size increased, while algae and terrestrial plant material became proportionally more important food resources downstream.

Food resources can be more broadly classified as autochthonous (i.e. algae, aquatic invertebrates, fish) or allochthonous (i.e. terrestrial invertebrates, terrestrial plant material). Regression analyses indicated that the relative importance of allochthonous foods increased slightly but significantly ($p = 0.016$) with stream size. A regional effect was also significant ($p = 0.040$) for the relative importance of aquatic versus terrestrial foods. Seven fish species found in Region 2, but not in Region 3, specialized on aquatic foods. This effect suggests that for a given stream size, the effectiveness of canopy shading is correlated with the proportion of the fish biomass supported by terrestrial foods.

Size distribution of fish. – Most individual fishes in our study streams were less than 90 mm TL. We analyzed the distributions of particularly small (<40 mm TL) and large (>100 mm TL) fish using multiple regression techniques to ascertain differences in distribution by size groups. Data from the R. Frijoles and R. Mendosa were not available for this analysis. Abundances (number m^{-2}) of small fish were correlated with habitat type, stream width and month (Fig. 10). Small fish were most common in deep pools and least common in riffles, a pattern observed for most feeding guilds. Small fish became less abundant in larger streams despite the fact that deep pools were more common there. Furthermore, most small fishes were general insectivores, which did not become less abundant with respect to biomass as stream size increased (see Guild distributions). Thus, small fishes were most common in deep pools of small streams. Small fishes were less abundant in the R. Limbo than expected for a stream its size (Fig. 10), again illustrating the impact of severe seasonality in discharge. Small fishes were also more abundant in March than January (Fig. 10) probably a result of reduced habitat availability at the end of the dry season. A similar pattern of seasonality was observed for numbers of fish >100 mm TL ($p = 0.005$).

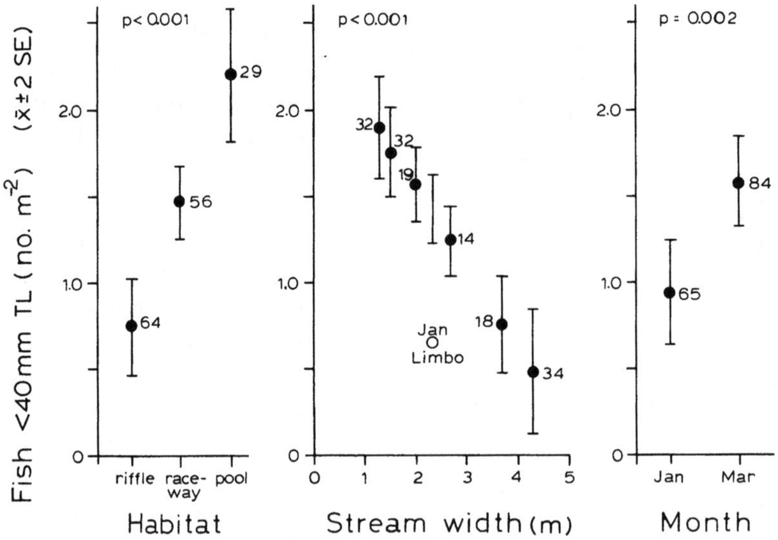

Fig. 10. Densities of small fish (< 40 mm TL) as a function of habitat type, stream size, and month. Numbers beside means indicate sample sizes associated with values taken by independent variables (horizontal axes). Means and standard errors are estimated from a multiple regression model.

Discussion

Distributions of animals among available habitats are generally mediated by 3 factors: food availability, predation intensity and tolerance of physicochemical conditions (Connell 1975). Commonly, these factors impose conflicting selection pressures on an individual's habitat use and, as a result, adaptations evolve to reconcile opposing selective forces. In fishes and other vertebrates, which are characterized by complex sensory capabilities and behavioral responses, habitat selection is an important adaptive component of a species' ecology (Partridge 1978, Morse 1980). Habitat selection by fishes may vary with age, sex, reproductive state, geographic area, and environmental conditions (Karr et al. 1982), and represents an integration of numerous ecological factors. The fluvial processes responsible for forming pools and riffles, however, are similar in all streams, and provide a basis for comparing patterns of habitat use among streams.

The impacts of physicochemical stresses on fish communities of tropical streams are most apparent in waters with exceptionally low pH (Roberts 1973, Janzen 1974, Lowe-McConnell 1975) or strongly seasonal discharge (Lowe-McConnell 1975, Goulding 1980). To our knowledge, pH, temperature, dissolved oxygen, or other physicochemical features of streams along the Pipeline Road are not particularly stressful to the fishes living there. Thus, we interpret observed differences in fish distribution among habitats and streams only in light of distributions of food resources and predators since these two factors are probably the primary determinants of habitat selection by these fishes.

Food availability patterns

Benthic invertebrates. – A major food resource used by stream fishes is benthic invertebrates. Benthic invertebrate abundances in this study varied among habitats, streams and seasons (Fig. 3). Other workers have reported invertebrates to be more abundant in riffles than in pools for both temperate (Rabeni & Minshall 1977, Slobodchikoff & Parrott 1977) and tropical (Fittkau 1964, Petr 1970, Bishop 1973) streams. Temperate populations of invertebrates undergo marked seasonal fluctuations in abundance, perhaps in response to pulses in availability of the organic material on which they feed (Hynes 1970). Lowest invertebrate numbers usually occur during late summer when discharges are also likely to be lowest, a pattern analogous to that reported here for tropical streams. A study by Power (1983) on the R. Frijoles indicates that algal productivity is low in the dry season despite an increase in light availability. This result is consistent with the seasonal patterns we observed for benthic invertebrates, many of which rely on algae for food.

The differences in benthic invertebrate abundances we observed among regions along the Pipeline Road may be explained on the basis of presumed productivity differences among regions. Increasing abundances through Regions 1, 2 and 3 are consistent with patterns expected if light limits primary productivity, which, in turn, limits secondary productivity. Reduced invertebrate abundances in tropical streams due to scouring discharge have been recently reported by Stout (1982) but since dry season rainfall differs little among Regions 1, 2 and 3 (Panama Canal Co. climatological data), differences in benthos abundances among regions are probably not due to differences in scouring frequency.

Terrestrial invertebrates. – The importance of terrestrial invertebrates as a food resource for stream fishes is particularly apparent in tropical systems (Inger & Chin 1962, Lowe-McConnell 1975). Yet, to our knowledge, no data on temporal and spatial variability of this resource are available for comparison. Anecdotal reports (Zaret & Rand 1971, Lowe-McConnell 1975) support the pattern observed here (Fig. 4) of lower abundances during the dry season. Our results are also consistent with data from Barro Colorado Island where canopy insects >5 mm were more numerous in January than March for 4 of the 5 years from 1972–1976 (Rubinoff 1974, Windsor 1975, 1976, 1977). Terrestrial invertebrates may usually be more available to stream fish in wet seasons due to higher general productivity and greater stream area relative to dry seasons. Furthermore, rain and runoff are probably effective means of transferring invertebrates from their normal substrates to streams.

Spatial variation in terrestrial invertebrate abundances is probably more complex than seasonal variation. One might expect terrestrial invertebrates to be most abundant where foliage is most dense. Our data indicate, however, that wide streams (effectively more open canopies) have greater numbers of terrestrial invertebrates than narrow ones (Fig. 4). The causal mechanisms for this pattern are unknown, but may be related to changes in foliage density profiles or air turbulence along the canopy openness gradient.

Other food resources. – Data on the temporal and spatial dynamics of other major food resources used by stream fish are rarely seen in the literature, yet certain patterns are expected. For example, the availability of attached algae to fish and invertebrates should increase as the canopy opens due to higher light availability, provided that stable substrates are available. However, reductions in algal availability during the dry season were reported by Power (1983) as a result of siltation and a decline in the amount of submerged substrate. Other potential food resources such as terrestrial fruits, flowers, and green leaves are probably also available seasonally due to pulses in production. Seasonal fluctuations in the availability of these foods are pronounced in South American forests (Gottsberger 1978, Goulding 1980) and are accentuated by seasonal changes in fishes' accessibility to them. Fishes relying on fruits and seeds for food have evolved seasonal migratory behaviors to effectively exploit this resource when the forest is inundated (Gottsberger 1978, Goulding 1980).

Food use by fishes

Dietary shifts with maturity. – Fish diets represent an integration of food preferences with food availability and accessibility. Changes in diet with fish size were apparent in: *Aequidens*, *Astyanax*, *Pimelodella*, and *Brycon* (Table 6). In all 4 species diets of small individuals were predominantly composed of insects, while larger individuals increased their intake of other food types. In *Brycon* the largest individuals specialized on terrestrial plant material. It is unknown whether these dietary shifts resulted from changes in food preferences or changes in the abilities of fish to acquire preferred foods. For example, larger fishes may be more adept, because of their greater speed, at acquiring food items striking the surface than smaller fishes. Certainly, larger fish are morphologically capable of using a wider range of food sizes. Changes in diet with fish size have been reported for numerous North American species that grow relatively large (Pflieger 1975, Keast 1977a, 1977b, Smith 1979). In a study of Amazonian fishes Knoppel (1970) concluded that dietary differences among conspecifics of different sizes were not substantial, although 4 of the 7 species he examined increased the proportion of terrestrial insects eaten as they matured, and 5 of the 7 species decreased the proportion of aquatic invertebrates eaten as they matured. The degree of change in food habits with fish size may be related to the relative importance of intra- and interspecific competition for food (Keast 1977b).

Fish distributions

Food and predator abundances in this study varied along gradients of habitat type, stream size and region. We examined these distributions with respect to fish distributions in an attempt to estimate the relative importance of food availability and predation to the organization of fish communities.

Habitat gradients. – Aquatic invertebrates were more abundant in riffles than pools (Fig. 3), yet aquatic insectivore fishes were not particularly concentrated into riffles. However, some benthic genera (e.g. *Imparales*, *Trichomycterus*) were found almost exclusively in riffles. Terrestrial foods are presumably equally abundant in all habitats, yet fishes relying on those foods were most concentrated into pools. The relative paucity of fish in riffles might be explained on the basis of the energetic expense incurred by a fish trying to maintain its position in fast turbulent water. However, this argument does not account for the differences in fish densities observed between raceways and pools (Fig. 6), where currents were negligible. A simpler explanation for the concentration of fishes into deep habitats is that deep water affords refuge from numerous avian and mammalian predators that inhabit the area (4 species of kingfishers, several species of herons and numerous mammals). Power (1983) presents evidence for the importance of predation pressure as a determinant of loricariid distribution in the R. Frijoles. Additional support comes from the low densities of fishes observed in the highly seasonal R. Limbo where predation by birds and mammals is probably particularly intense during dry periods.

A trade-off exists for small fishes that enter pools to avoid terrestrial predators in that piscine predators (generally relatively large fishes) are more common in deep habitats. Small fishes could concentrate into pools of smaller streams where piscine predators are not very common. Our results are consistent with this hypothesis, but the majority of small fishes in our study streams were poeciliids and juvenile characoids, mostly general insectivores. Since their numbers declined significantly downstream (Fig. 9) but general insectivore biomass did not, small fishes seemed to be replaced by larger characoids downstream. Further study is necessary to distinguish whether small general insectivores concentrate into upstream pools to avoid larger competitors or piscine predators.

The concentration of small fish into pool habitats in Panama streams is in contrast with the pattern observed by Schlosser (1982) in a similar size stream in Illinois. In Illinois, young-of-the-year cyprinids and centrarchids, which feed largely on aquatic invertebrates, were most abundant in riffle habitats. Similarly, Power (1983) reported that juvenile loricariid catfishes (algae grazers) were more abundant in riffles, where algal densities were greatest.

Gradients among streams. – Aquatic invertebrates were more abundant in less shaded streams (Fig. 3) while terrestrial invertebrates were more abundant in larger streams (Fig. 4). Neither aquatic insectivore nor general insectivore densities were closely correlated with abundances of their food resources among streams. In contrast, algivore densities increased downstream, a pattern that might also be expected for abundance of attached algae since light availability increases downstream. *Hoplias* (piscivore) was never found in Region 3 and *Roeboides* (scale-eater) was found in only 1 stream of Region 3. Since fishes in general were least abundant in Region 3, the distributions of these 2 fish-exploiters seem to correspond to the distribution of their food resources. Unfortunately, we have no data on variation in abundances of fish-eating birds among different streams and regions. We assume, therefore, that they exert similar fishing pressures on all streams.

We conclude that fish biomass distributions among habitats within a stream appear to be more closely related to dangers of predation by birds and mammals than to the distribution of food resources. Distributions of small fish, however, are more closely correlated with food abundance. Distributions of fish biomass along gradients of stream size and canopy openness are complex, the correlations with food abundances varying among feeding guilds.

Relative importance of feeding guilds. – Streams along the Pipeline Road were dominated (both biomass and numbers) by fishes with relatively generalized diets (general insectivores and omnivores). Fishes with more specialized diets (algivores, terrestrial herbivores, piscivores) generally comprised minor components of the fish community, but often became relatively more abundant in larger streams (Fig. 8). In addition, species richness of feeding guilds increased with stream size and canopy openness (Fig. 7). At the community level, biomass of fish dependent on autochthonous food resources (i.e. algae, benthos, fish) decreased as stream size increased but increased as canopy openness increased for a given stream size.

These patterns suggest that increases in trophic diversity may be generally correlated with increases in food resource 'reliability', where reliability is some measure of productivity and predictability. This hypothesis, supported by a theoretical model of species packing developed by MacArthur (1970), can be used to explain differences in stream trophic diversity between such widely separated regions as Panama and the midwest United States. Midwestern U.S. streams typically support 1 or 2 algivorous species (Pflieger 1975, Smith 1979) compared with 5 algivores in the R. Frijoles. In addition, no fish species in the Midwest specialize on terrestrial plant material as do adult *Brycon* in Panama. The relative importance of terrestrial invertebrates as a food resource is typically less in the Midwest than in Panama (Lotrich 1973, Pflieger 1975, Smith 1979, Angermeier 1982). Parallel explanations of trophic diversity have been proposed to account for differences in feeding guild structure between temperate and tropical communities of forest birds (Karr 1975) and bats (Fleming 1973). Temperate system patterns similar to those observed here for Panama include downstream additions of species and guilds (Kuehne 1962, Sheldon 1968, Lotrich 1973, Gorman & Karr 1978, Horwitz 1978). (In contrast to our observations, Lotrich 1973, reported a decline in the importance of allochthonous foods to the fish community as stream size increased in Kentucky.)

Our work in small streams in Panama, as well as studies by others in large tropical rivers (Lowe-McConnell 1975, Goulding 1980), indicates that tropical fishes are dependent on foods derived directly from the riparian forest. This pattern suggests that large scale alterations in forest composition and structure may have serious impacts on the integrity of tropical stream communities. Deforestation, pollution and other human disturbances are likely to impose major effects on shading, discharge variability, siltation, nutrient loads and availability of various foods. Precise predictions of how these environmental changes will affect fish community structure are not possible, in part, because little is known of the natural history of most tropical species, but analogous modifications in Midwest streams have resulted in lower species and trophic diversities and higher rates of disease and hybridization (Greenfield et al. 1973, Karr 1981). Indirect effects of deforestation on Midwest streams include elevated and more variable water temperatures, lower discharge during drought periods, and shifts in habitats toward shallower pools and finer substrates. These conditions select for species that are more tolerant of physicochemical stress, generalized in food habits, and lack strict spawning requirements. Species that depend directly on allochthonous foods such as fruits may be lost since such resources probably become more scarce. Increases in siltation would likely inhibit autochthonous productivity, perhaps resulting in fewer fish and invertebrates that require algae for food. Loss of deep pools through siltation or channelization could shift the size structure of the fish community toward small individuals. Clearly, more information on the requirements and interactions of stream organisms is necessary if the integrity of tropical streams is to be maintained in the face of expanding human populations and technology.

Acknowledgements

Portions of this work were supported by USEPA (R806391) and Earthwatch grants to JRK and University of Illinois Research Board and Sigma Xi grants to PLA. We thank RENARE for permitting us to work in Parque Nacional Soberania. We are grateful for assistance from T.E. Martin, S. Moore, I.J. Schlosser and J.J. Vizek in the field and from K.A. Bisbee with data analysis. K. Fauch, M. Power, S. Rand, M. Robinson, I. Schlosser, P. Yant, T. Zaret and anonymous reviewers made numerous insightful comments on an earlier draft of the manuscript. J. Lunberg helped clarify several questions on systematics.

References cited

Angermeier, P.L. 1982. Resource seasonality and fish diets in an Illinois stream. Env. Biol. Fish. 7: 251–264.

Bishop, J.E. 1973. Limnology of a small Malayan river Sungai Gombak. Monogr. Biol. 22. Dr W. Junk Publishers, The Hague. 485 pp.

Connell, J.H. 1975. Some mechanisms producing structure in natural communities: a model and evidence from field experiments. pp. 460–490. *In*: M.L. Cody & J.M. Diamond (ed.) Ecology and Evolution of Communities, Belknap Press, Cambridge.

Fittkau, E.J. 1964. Remarks on limnology of central-Amazon rain-forest streams. Verh. int. Verein. theor. angew. Limnol. 15: 1092–1096.

Fleming, T.H. 1973. Numbers of mammal species in North and Central American forest communities. Ecology 54: 555–563.

Gorman, O.T. & J.R. Karr. 1978. Habitat structure and stream fish communities. Ecology 59: 507–515.

Gottsberger, G. 1978. Seed dispersal by fish in the inundated regions of Humaitá Amazonia. Biotropica 10: 170–183.

Goulding, M. 1980. The fishes and the forest: explorations in Amazonian natural history. University of California Press, Berkeley. 280 pp.

Greenfield, D.W., F. Abdel-Hameed, G.D. Deckert & R.R. Flinn. 1973. Hybridization between *Chrosomus erythrogaster* and *Notropis cornutus* (Pisces: Cyprinidae). Copeia 1973: 54–60.

Horn, H.S. 1966. Measurement of overlap in comparative ecological studies. Amer. Nat. 100: 419–424.

Horwitz, R.J. 1978. Temporal variability patterns and the distributional patterns of stream fishes. Ecol. Monogr. 48: 307–321.

Hynes, H.B.N. 1970. The ecology of running waters. University of Toronto Press, Toronto. 555 pp.

Inger, R.F. & P.K. Chin. 1962. The freshwater fishes of North Borneo. Fieldiana (Zool.) 45: 1–268.

Janzen, D.H. 1974. Tropical blackwater rivers, animals, and mast fruiting by the Dipterocarpaceae. Biotropica 6: 69–103.

Karr, J.R. 1975. Production, energy pathways and community diversity in forest birds. pp. 161–176. *In*: F.B. Golley & E. Medina (ed.) Tropical Ecological Systems: Trends in Terrestrial and Aquatic Research, Ecological Studies Vol. II, Springer-Verlag, New York.

Karr, J.R. 1981. Assessment of biotic integrity using fish communities. Fisheries 6: 21–27.

Karr, J.R. & L.A. Dudley. 1981. Ecological perspective on water quality goals. Envir. Manage 5: 55–68.

Karr, J.R., L.A. Toth & G.D. Garman. 1982. Habitat preservation for midwest stream fishes: principles and guidelines. U.S. Environmental Protection Agency 600/3-83-006, Corvallis. 120 pp.

Keast, A. 1977a. Diet overlaps and feeding relationships between the year classes in the yellow perch (*Perca flavescens*). Env. Biol. Fish. 2: 53–70.

Keast, A. 1977b. Mechanisms expanding niche width and minimizing intraspecific competition in two centrarchid fishes. Evol. Biol. 10: 333–395.

Knöppel, H.A. 1970. Food of central Amazonian fishes: contribution to the nutrient ecology of Amazonian rain forest streams. Amazoniana 2: 257–352.

Kuehne, R.A. 1962. A classification of streams illustrated by fish distribution in an eastern Kentucky creek. Ecology 43: 608–614.

Lotrich, V.A. 1973. Growth, production and community composition of fishes inhabiting a first-, second-, and third-order stream of eastern Kentucky. Ecol. Monogr. 43: 377–397.

Lowe-McConnell, R.H. 1975. Fish communities in tropical freshwater: their distribution, ecology and evolution. Longman, London. 337 pp.

MacArthur, R.H. 1970. Species packing and competitive equilibrium for many species. Theor. Pop. Biol. 1: 1–11.

Morse, D.H. 1980. Behavioral mechanisms in ecology. Harvard University Press, Cambridge. 383 pp.

Parker, R.R. 1963. Effects of formalin on length and weight of fishes. J Fish. Res. Board Can. 20: 1441–1445.

Partridge, L. 1978. Habitat selection. pp. 351–376. *In*: J.R. Krebs & N.B. Davies (ed.) Behavioral Ecology: An Evolutionary

Approach. Blackwell Scientific Publications, Oxford.
Petr, T. 1970. The bottom fauna of the rapids of the Black Volta River in Ghana. Hydrobiologia 36: 399–418.
Pflieger, W.L. 1975. The fishes of Missouri. Missouri Dept. of Conservation, Jefferson City. 343 pp.
Power, M.E. 1983. Grazing ecology of loricariid catfishes in a Panamanian stream. Env. Biol. Fish (In press).
Rabeni, C.F. & G.W. Minshall. 1977. Factors affecting microdistribution of stream benthic insects. Oikos 29: 33–43.
Roberts, T.R. 1973. Ecology of fishes in the Amazon and Congo basin. pp. 239–254. *In*: B.J. Meggers, E.S. Eyensu & W.D. Duckworth (ed.) Tropical Forest Ecosystems in Africa and South America: A Comparative Review. Smithsonian Inst. Press, Washington.
Rubinoff, R.W. (ed.) 1974. Environmental monitoring and baseline data – 1973. Smithsonian Institution Environmental Sciences Program. Smithsonian Inst., Washington. 45 pp.
Schlosser, I.J. 1982. Fish community structure and function along two habitat gradients in a headwater stream. Ecol. Monogr. 52: 395–414.
Sheldon, A.L. 1968. Species diversity and longitudinal succession in stream fishes. Ecology 49: 193–198.
Slobodchikoff, C.N. & J.E. Parrott. 1977. Seasonal diversity in aquatic insect communities in an all-year stream system. Hydrobiologia 52: 143–151.
Smith, P.W. 1979. The fishes of Illinois. University of Illinois Press, Urbana. 314 pp.
Sokal, R.R. & P.H.A. Sneath. 1963. Principles of numerical taxonomy. W.H. Freeman, San Francisco. 359 pp.
Stout, R.J. 1982. Effects of a harsh environment on the life history patterns of two species of tropical aquatic hemiptera (Family: Naucoridae). Ecology 63: 75–83.
Vannote, R.L., G.W. Minshall, K.W. Cummins, J.R. Sedell & C.E. Cushing. 1980. The river continuum concept. Can. J. Fish. Aquat. Sci. 37: 130–137.
Waters, T.F. 1973. Subsampler for dividing large samples of stream invertebrate drift. Limnol. and Oceanogr. 18: 813–815.
Windsor, D.M. (ed.) 1975. Environmental monitoring and baseline data – 1974. Smithsonian Institution Environmental Sciences Program. Smithsonian Inst., Washington. 409 pp.
Windsor, D.M. (ed.). 1976. Environmental monitoring and baseline data – 1975. Smithsonian Institution Environmental Sciences Program. Smithsonian Inst., Washington. 252 pp.
Windsor, D.M. (ed.). 1977. Environmental monitoring and baseline data from the isthmus of Panama – 1976. Smithsonian Institution Environmental Sciences Program. Smithsonian Ins., Washington. 267 pp.
Zaret, T.M. & A.S. Rand. 1971. Competition in tropical stream fishes: support for the competitive exclusion principle. Ecology 52: 336–342.

Originally published in Env. Biol. Fish. 9: 117–135

Male of *Parauchenipterus insignis* from Steindachner (1879), Zur Fisch-Fauna des Magdalenen-Stroms. Denkschr. Akad. Wiss. Wien 39: 19–78.

Detritivory in neotropical fish communities

Stephen H. Bowen
Department of Biological Sciences, Michigan Technological University, Houghton, MI 49931, U.S.A.

Keywords: *Curimatus*, Detritus, Diet, Digestion, Feeding behavior, Strategies, Morphology, *Prochilodus*

Synopsis

Fish communities of major river systems in South America contain a high proportion of detritivorous fishes in the families Prochilodontidae and Curimatidae. These families include important fish stocks that in some regions comprise over 50 percent of the community ichthyomass. As a group, detritivores have anatomical-physiological adaptations for collection and digestion of detritus, but the actual mechanisms of these presumed adaptations have to-date only been inferred. Dietary requirements have not been identified. Behavioral adaptation is implied by feeding habitat selection but its nutritional significance is unknown. Because many of these species have commercial importance, and because ongoing construction of impoundments threatens to disrupt seasonal migrations between spawning and feeding areas, an understanding of the feeding biology of detritivores is important.

The number and distribution of neotropical detritivorous fish

A fundamental discovery of ecological research during the last 15 years is that a principal route of ecosystem energy flux and material cycling is through the detritus food chain. Although herbivores are conspicuous in their consumption of primary food resources, inconspicuous soil and sediment detritivores do much of the work in conversion of plant matter to animal biomass. The relative importance of detritivory varies from system to system, but estimates for some ecosystems attribute 90% or more of primary consumption to detritivores (Mann 1972).

Only a small percentage of fish species feed as detritivores (Lowe-McConnell 1975). Although many fishes ingest small quantities of detritus[1] incidentally while feeding on benthic prey, this has not been considered to be nutritionally significant. Other fishes may turn to detritus as a temporary diet when preferred foods are not available, but their rapid loss of condition during this period indicates they are ill-prepared to exploit a detritus diet (Lowe-McConnell 1975). The great majority of fishes feed as secondary or higher level consumers and rely on invertebrates as their link to the detritus base of the food chain (Eggers et al. 1978).

The exception to this rule is found in the tropics, where detritivorous fishes can dominate ecosystem ichthyomass. This is most evident in the neotropics,

[1] The term detritus will be used to mean dead organic matter that has been altered in some way that renders it unlike its original living form, i.e. organic matter that has undergone some diagenesis. By this definition, I exclude from consideration allochthonous fruits and flowers that are important in the diets of fishes feeding in seasonally flooded forests.

Thomas M. Zaret (ed.), Evolutionary ecology of neotropical freshwater fishes. ISBN 90 6193 823 6

especially in the great river systems of South America. Most of the abundant detritivorous fishes in these rivers are contained in two closely related families: Prochilodontidae and Curimatidae (Fig. 1). The Loricariidae are sometimes mentioned as possible detritivores, but recent data indicate they feed principally on algae (M.E. Power, this issue). Due to the vast dimensions and open nature of these riverine ecosystems, there are only a few quantitative estimates of the ichthyomass comprised by a single detritivorous species. For the Rio Paraná (Rio de la Plata system), Bonetto (1970) estimates from extensive survey data that *Prochilodus platensis* compromises 60% of the total ichthyomass. In the Rio Pilcomayo (Rio de la Plata system), *P. platensis* also makes up 'the major part of the fish biomass' (Bayley 1973). In the Rio Madeira (Amazon system), the nine species of Prochilondontidae and Curimatidae together appear to be similarly dominant (Goulding 1981). The distribution of neotropical detritivores extends far upstream, although they may diminish in importance in lower-order streams (Lowe-McConnell 1975, p. 208). They are abundant in floodplain pools (Bonnetto et al. 1969) and were once abundant in one of the few natural low-altitude lakes in South America (Lake Valencia, Venezuela) (Pearse 1920). In view of their abundance, these few detritivorous fishes must play major roles in ecosystem energy flux and material cycling and in the population dynamics of their respective fish communities.

In Africa, there are three groups of detritus feeding fishes: the Citharinidae, the Cyprinidae including *Labeo* sp., and some of the Cichlidae. Citharinids are closely related to neotropical detritivores (same suborder Characoidei, Nelson 1976), and they occupy identical habitats in low-gradient rivers, backwaters and floodplain pools. The cyprinid *Labeo* sp. occupies both riverine and lacustrine habitats in Africa, and the family Cyprinidae is entirely absent from the neotropics. Cichlids are widespread in both neotropical and African regions, but only in Africa do they appear as successful detritivores in shallow lakes and floodplain pools (Bowen 1979a). The detritivorous African cichlid *Sarotherodon mossambicus* has been introduced to freshwaters in much of the neotropical region, and in some areas large wild populations have become well established (Bowen 1980). Thus,

Fig. 1. Prochilodus platensis from the Riachuelo backwater, Rio Parana near Corrientes, Argentina.

it remains a mystery why neotropical cichlids have not evolved to take advantage of the detritus food resource.

For other regions, there is too little information to make useful comparisons. Detritus is clearly present in the diet of many temperate fishes, especially cyprinids, freshwater clupeids and catostomids. It will be essential to quantify the role of detritus in the nutrition of these groups before the trophic identification is certain.

Adaptations to detritivory

Studies of *S. mossambicus* indicate that two types of adaptations are important to the success of detritivorous fishes: morphological adaptations of the digestive system and behavioral adaptations for location of detritus (Bowen 1979a, 1979b, 1980, 1981). Observations reported for neotropical detritivores suggest similar adaptations.

The alimentary canal

The structure of the alimentary canal in *Prochilodus platensis* from the Rio de la Plata is described in detail by Angelescu & Gneri (1949). I found *Semaprochilodus* sp. and *Curimatus* sp. to have similar digestive tracts, although the relative dimensions of various structures may vary. Thus, general features of the digestive tract in *P. platensis* described by Angelescu & Gneri may be taken as representative of all Prochilodontidae and Curimatidae, although the function of this digestive tract has not been described.

In *P. platensis*, the protrusible jaws are weak and the lips bear fine delicate teeth. Observed in aquaria, the fish suck in fine, flocculant detritus from the surfaces of vascular plants and from the bottom. The teeth do not play a discernable role in this behavior.

Many detritivorous fishes confront difficulties associated with large quantities of inorganic matter, often mostly sand, mixed with organic detritus. When feeding from the bottom, both *S. mossambicus* and the ubiquitous estuarine and marine detritivore *Mugil cephalus* sort sediment in their oral cavity and reject inorganic matter while they retain organic detritus (Odum 1968, Bowen 1979a). The unusual oral cavity in *P. platensis* may improve the ability of the fishes to sort. Viewed in transverse section, the oral cavity has an inverted V shape (Fig. 2). In the anterior half, a single median dorsal, paired dorsolateral and paired lateral ridges extend into the dorsal portion of the cavity (Fig. 3a). The ridge margins are papillose, suggesting a sensory function. As sediment is sucked into the oral cavity, the coarse inorganic particles would be expected to settle into the lower arms of the inverted V, while flocculant organic matter would stay suspended. The five oral ridges could facilitate this

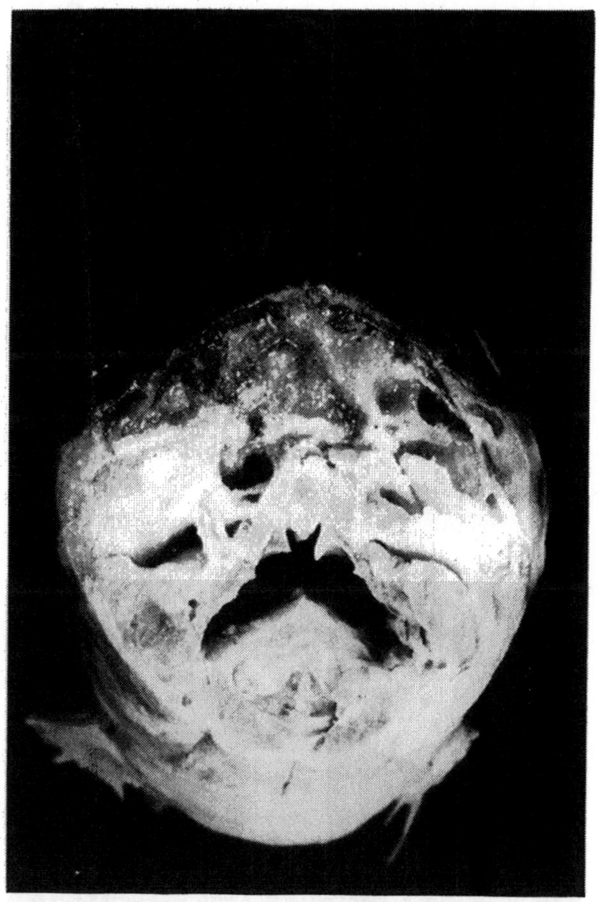

Fig. 2. Transverse section through the posterior oral cavity of *Prochilodus platensis*. Note the inverted V shape of the cavity and the ridges in the dorsal portion.

separation by reduction of turbulence, and the sensory structures assess the quality of the suspended organic matter.

Separation of flocculant detritus from water is probably achieved by the gillrakers. Unlike many fishes in which gillrakers form an essentially two-dimensional net, rakers on dorsal and ventral branches of the acutely bent gill arches intermesh to form a three-dimensional filtering structure. The individual rakers are fleshy, elongate structures oriented at an oblique angle to the gill arch. As suspended flocculant detritus is forced into this maze of constricted passageways, its adhesive nature, mucus on the rakers and the force of water flowing toward the aboral cavity would be expected to aggregate the detritus in a form that could be moved down the esophagus by peristalsis.

Epibranchial organs are present in the posterior pharynx (Fig. 3b). Similar structures are present in many lower teleosts, and their function appears to vary from species to species (Kapoor et al. 1975). Angelescu & Gneri (1949) suggested they produced mucus to aid passage of food through the esophagus, but stomach contents of freshly caught *P. platensis* I have examined do not include noticeably more mucus than those of *S. mossambicus*, which feed on a nearly identical diet but lack epibranchial organs. The food of *P. platensis* does enter the epibranchials, and thus their role in digestion merits closer scrutiny.

The stomach of *P. platensis* is divided into two parts (Fig. 4). The more anterior cardiac stomach is

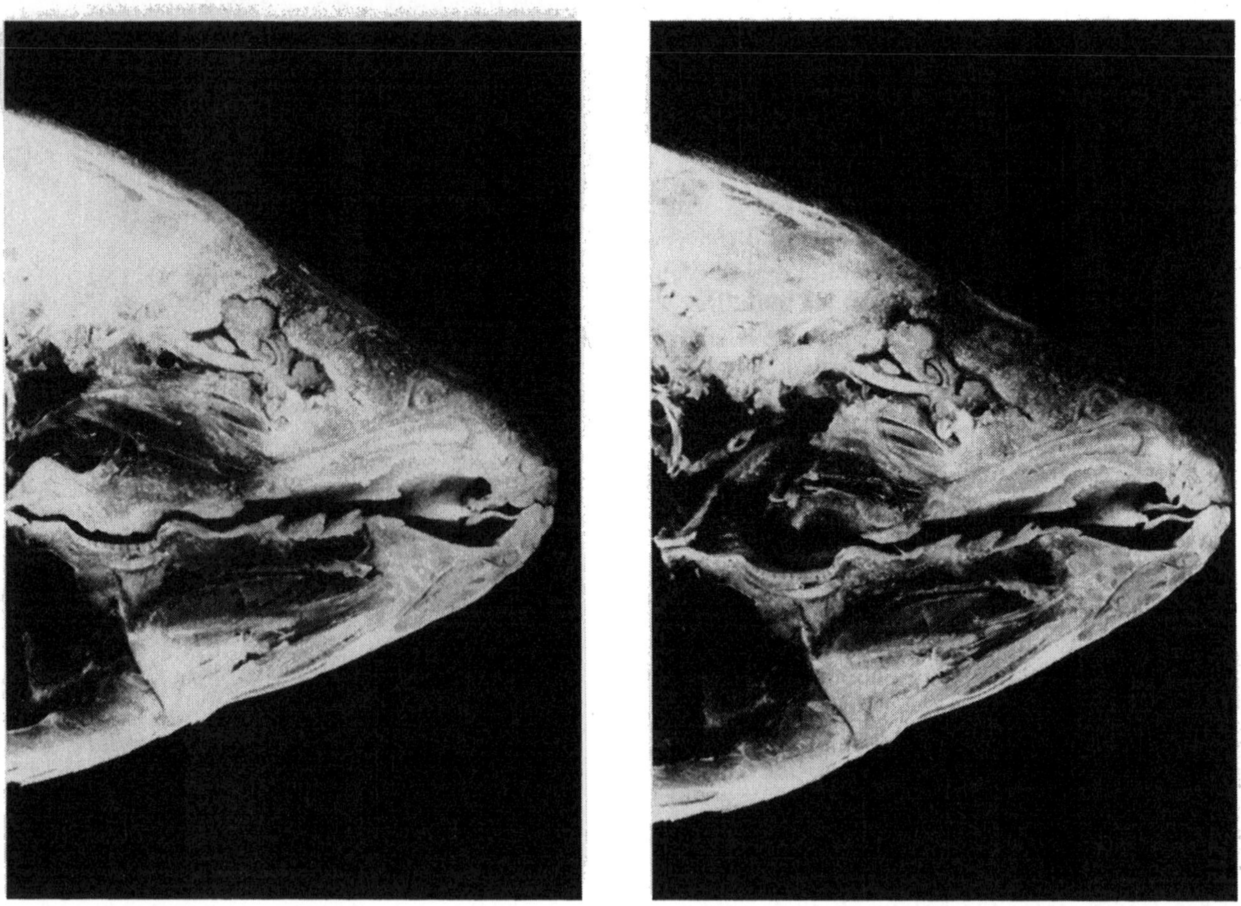

Fig. 3. Sagittal section of *Prochilodus platensis* head: in a – note the development and papillose margins of the oral ridges; in b – the dorsal pharyngeal pad has been removed to expose the left epibranchial organ.

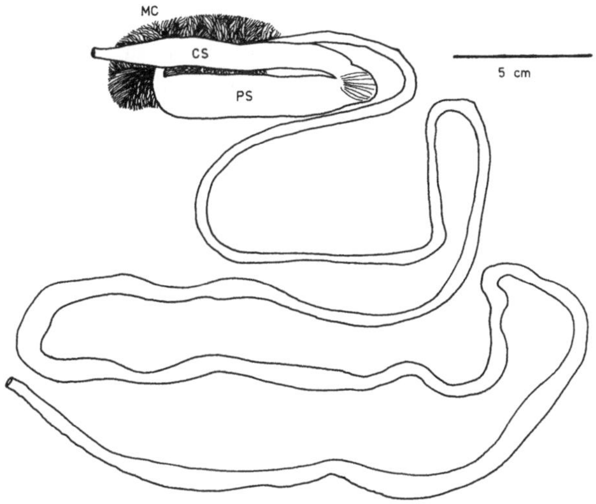

Fig. 4. Dissected digestive tract of a 31 cm (SL) *Prochilodus platensis.* (MC = mass of caeca, CS = cardiac stomach, PS = pyloric stomach.) In undissected specimens, anterior pyloric stomach is surrounded by the caecal mass. Traced from a photographic projection.

muscular but distensible so that it acts as a reservoir. The pyloric stomach is extremely muscular and nearly rigid, and serves to grind the food together with sand also ingested. This reduces the detritus to a small, nearly uniform particle size, conducive to enzyme-substrate interaction, and to peristalsis in the intestine (Fig. 5).

Similar grinding stomachs are described for *Citharinus* sp. and *Mugil* sp. (Kapoor et al. 1975), and Payne (1978) has suggested that gastric grinding in *Mugil* serves to disrupt algal cell walls. Since algae are rare in the diet of *P. platensis*, cell wall disruption is unlikely to be important for this species (Bowen personal observation). In sharp contrast to *Prochilodus*, *Citharinus* and *Mugil*, the stomachs in detritivorous cichlids are thin-walled, blind sacs. Many *Tilapia* and *Sarotherodon* sp. secrete gastric acid to unusually low pH values, frequently below pH 1.5. These conditions both disrupt cell walls (Moriarty 1973, Bowen 1976, Payne 1978) and fundamentally alter the chemical structure of detritus in ways that may facilitate intestinal digestion (Bowen 1981). Detritivores in the genus *Labeo* differ from others in that they lack a stomach and thus digestion in these fish appears to be a one-step, intestinal process.

Between the pyloric stomach and the intestine is a structure termed the pyloric chamber that gives rise laterally to short-branched second- and third-order chambers which finally branch into a mass of villiform pyloric caeca (Fig. 4). The caeca are approximately 1 mm in diameter and range from 3 to 10 mm in length. Angelescu & Gneri (1949) counted 3000 caeca in one specimen. Food from the

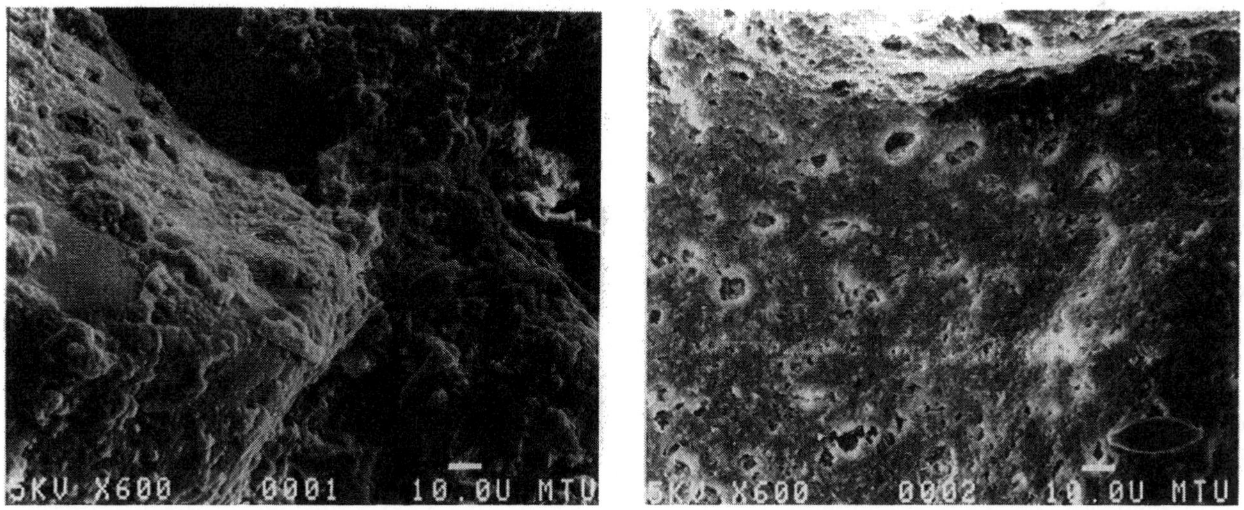

Fig. 5. Stomach contents of *Prochilodus platensis* from the Riachuelo backwater near Corrientes, Argentina: a – food from the cardiac stomach varies greatly in particle size; b – after grinding in the pyloric stomach, particle size is uniformly small.

pyloric stomach passes into the pyloric chamber, enters the caeca and is later extruded back into the pyloric chamber from which it passes on to the intestine. Food in the anterior intestine is often in the form of tiny strands with dimensions of the pyloric caecum lumen. From histological data, Domitrovic (personal communication) concluded the pyloric caeca are not secretory, but serve in post gastric assimilation. This differs from the largely secretory role ascribed to pyloric caeca in other fishes (Kapoor et al. 1975).

Although no sphincter separates the pyloric chamber from the intestine in *P. platensis*, the functional beginning of the intestine is found at the point immediately posterior to the pyloric chamber where the common bile duct delivers bile salts and intestinal digestive enzymes to the lumen of the gut. Two regions of the intestine are readily distinguished according to the type of folds formed by the intestinal mucosa. Folds in the upper third are circular and low in profile. In the lower two-thirds, partition-like, transverse folds that extend halfway across the lumen are found on alternate sides of the intestine space at 3 to 8 mm intervals.

Among fishes, the degree of intestinal development in length and mucosal folding is clearly correlated with trophic status, ranked according to relative intestinal development, carnivores < omnivores < herbivores < detritivores (Fryer & Iles 1972, Kapoor et al. 1975). This relationship is generally interpreted as a reflection of the resistance of different foods to intestinal enzymic digestion. Animal prey that have been masticated or processed in the stomach offer few barriers to intestinal enzymes. In contrast, plant and bacterial cell walls are very effective barriers. Even with well developed means for cell-wall disruption, intestinal digestion appears to be much slower in herbivores than in carnivores (Moriarty 1973, Bowen 1976, Caulton 1976).

A different sort of barrier may retard intestinal digestion of detritus. During the process of detritus formation, the chemical structure of the parent material is extensively modified. Many reactive sites previously available to digestive enzymes may be altered, or masked by other organic or inorganic compounds, such that few sites are available for enzymic hydrolysis. A study of intestinal digestion in *S. mossambicus* showed that the full length of the long intestine was necessary for maximum digestion of detrital amino acids (Bowen 1980). Had these amino acids been present in the form of protein (parent material), digestion would have been very rapid (Bowen 1981). It appears that detritivores confront special obstacles to intestinal enzymic digestion as a consequence of chemical changes that are integral to detritus formation, and a well developed intestine is needed to overcome these obstacles.

There are two ways in which the intestine may be developed. In *Labeo* and *Sarotherodon*, the intestine is exceptionally long: 15 to 21 and 8 to 10 times body length, respectively (Kapoor et al. 1975, Bowen 1982a, 1982b). In *Prochilodus*, *Curimatus* and *Citharinus* the intestine is relatively short (3 to 4 times body length), but the mucosal folds are extensively developed. Angelescu & Gneri (1949) estimate that mucosal folds increase the absorptive surface in the intestine of *P. platensis* by a factor of four. Thus, the available data indicate that the intestine is similarly well developed in each of these detritivores.

Although the above comparisons build an image of parallel adaptations developed to overcome common obstacles to digestion, truly definitive comparisons will require quantitative analysis of digestive processes.

Selective feeding

We can reasonably expect the flowing waters of neotropical rivers to provide a diverse menu of detrital food resources. Current velocity determines the size of detrital particles that settle out of suspension with fine particulate detritus accumulating only in the quiet backwaters. Watershed vegetation and water chemistry will determine the quality and quantity of dissolved organic matter that precipitates to form detritus in each tributary. As detritus is transported downstream, it is continuously modified by physical, chemical and biological agents that are likely to alter food quality.

Studies of *S. mossambicus* in Lake Sibaya, South

Africa, and Lake Valencia, Venezuela, have shown that selective feeding on high-quality detritus is the key to their success. In both lakes the amino acid content of diet varies across a wide range from low levels inadequate for simple body maintenance to high levels that support rapid growth (Bowen 1979, 1980). Those *S. mossambicus* that select high amino acid detritus grow rapidly and are in good condition. Apparently due to the conflicting demands of predator avoidance, adult *S. mossambicus* in Lake Sibaya feed on amino acid-poor detritus and suffer severe malnutrition as a result (Bowen 1979a). Similar work on *Citharinus citharinus* in backwaters of the Niger River has shown these fish select benthic detritus high in nitrogen (= amino acid?) (Bakare 1970).

Lowe-McConnell (1975) noted that, although organic detritus is abundant in all reaches, detritivorous fishes are generally abundant only in middle and lower reaches where fine-particulate detritus accumulates. Observations on the annual migratory cycles of several *Prochilodus* populations support her view. *Prochilodus* sp. ascend tributaries to spawn, usually during the rising flood cycle. Throughout the period of migration and spawning, they rely on large visceral fat reserves and do not feed (Lowe-McConnell 1975, Cannon personal communication). Once they have returned to downstream backwaters, feeding commences and the reserves are regenerated (Bayley 1973, Lowe-McConnell 1975, Cannon personal communication). Fishermen report that, within backwater areas, *Prochilodus* are commonly seen feeding in water less than 30 cm deep. A similar preference for detritus in shallow backwaters has been reported for mullet (Odum 1970). Is it possible that *P. platensis* select shallow backwater areas to feed because they contain detritus of superior food value?

Future research

In many parts of South America, detritivores are the principal species harvested in commercial fisheries (Mago 1972, Godoy cited in Lowe-McConnell 1975). In other areas, overharvest of predatory fishes may soon force the fishery to rely more heavily on detritivores (Goulding 1981). Increases in fishing pressure come at a time when these species face profound alterations of their natural environment by man. Several major impoundments are in various stages of planning and construction in the Orinoco and de la Plata basins. Some are planned without a fish-passage facility. For others the passage facility may not be properly designed. One species of *Semaprochilodus* is apparently unable to ascend rapids routinely negotiated by other fishes (Goulding 1981), and thus special design considerations may be required. Will species of Prochilodontidae and Curimatidae trapped below impoundments be able to spawn? Will those trapped above be able to find adequate feeding areas? What will be the effect of the increased lentic character of the system on the detritus food resource and the detritivore's ability to utilize it? These and other questions can only be answered by future research on detritivores.

Acknowledgements

I am grateful to Dr. A.A. Bonetto and his staff at the Centro de Ecologia aplicada del Litoral, Corrientes, Argentina, for their generous assistance in collection of fish samples examined for this report. Travel to Corrientes was supported by the International Programs Division, The National Sciences Foundation, U.S.A.

References cited

Angelescu, V. & F.S. Gneri. 1949. Adaptations of the alimentary canal in some species of the Uruguay river and La Plata river. I. Omnivores and iliovores in the families Loricariidae and Anostomidae. Rev. Inst. Nac. Cien. Nat. Cien. Zool. 1: 161–172. (In Spanish.)

Bakare, O. 1970. Bottom deposits as food of inland freshwater fish. pp. 65–85. *In*: S.A. Visser (ed.) Kainji Lake Studies, Univ. Press, Ibadan, Nigeria.

Bayley, P.B. 1973. Studies on the migratory characin, *Prochilodus platensis* Homberg 1889 (Pisces, Characoidei) in the River Pilcomayo, South America. J. Fish Biol. 5: 25–40.

Bonetto, A.A. 1970. Principal limnological features of northeastern Argentina. Soc. Arg. Bot. Supplement 11: 185–209. (In Spanish.)

Bonetto, A.A., W. Dioni & C. Pignalberi. 1969. Limnological investigations on biotic communities in the middle Paraná River Valley. Verh. intern. Verein. Limnol. 17: 1035–1050.

Bowen, S.H. 1976. Mechanisms for digestion of detrital bacteria by the cichlid fish *Sarotherodon mossambicus* (Peters). Nature 260: 137–138.

Bowen, S.H. 1979a. A nutritional constraint in detritivory by fishes: The stunted populations of *Sarotherodon mossambicus* in Lake Sibaya, South Africa. Ecolog. Monogr. 49: 17–31.

Bowen, S.H. 1979b. Determinants of the chemical composition of periphytic detrital aggregate in a tropical lake (Lake Valencia, Venezuela). Archiv Hydrobiol. 87: 166–177.

Bowen, S.H. 1980. Detrital nonprotein amino acids are the key to rapid growth of *Tilapia* in Lake Valencia, Venezuela. Science 207: 1216–1218.

Bowen, S.H. 1981. Digestion and assimilation of periphytic detrital aggregate by *Tilapia mossambicus*. Trans. Amer. Fish. Soc. 110: 241–247.

Bowen, S.H. 1982a. Feeding, digestion and growth-qualitative considerations. pp. 141–156. *In*: R.S.V. Pullin & R.H. Lowe-McConnell (ed.), The Biology and Culture of Tilapias, ICLARM Conf. Proc. 7, Manila.

Bowen, S.H. 1982b. Detritivory and herbivory. *In*: C. Leveque et al. (ed.) Biology and Ecology of African Freshwater Fishes (in press).

Caulton, M.S. 1976. The importance of pre-digestive food preparation to *Tilapia rendalli* when feeding on aquatic macrophytes. Trans. Rhod. Scien. Ass. 57: 22–28.

Eggers, D.M., N.W. Bartoo, N.A. Richard, R.E. Nelson, R.C. Wissmar, R.L. Burner & A.H. Devol. 1978. The Lake Washington ecosystem: The perspective from the fish community production forage base. J. Fish. Res. Board Can. 35: 1553–1571.

Fryer, G. & T.D. Iles. 1972. The cichlid fishes of the great lakes of Africa. Oliver and Boyd, Edinburg. 641 pp.

Goulding, M. 1981. Man and fisheries on an Amazon frontier. Developments in Hydrobiology 4, Dr W. Junk Publishers, The Hague, 137 pp.

Kapoor, B.G., H. Smit & I.A. Verighina. 1975. The alimentary canal and digestion in teleosts. Adv. in Mar. Biol. 13: 109–300.

Lowe-McConnell, R.H. 1975. Fish communities in tropical freshwaters. Longman, New York, 337 pp.

Mago, F.M.L. 1972. Considerations in the systematics of the family Prochilodontidae with a synopsis of the Venezuelan species. Acta Biol. Venez. 8: 35–96. (In Spanish.)

Mann, K.H. 1972. Introductory remarks. pp. 13–16. *In*: Melchiorri-Santolini & J.W. Hopton (ed.) Detritus and its Role in Aquatic Ecosystems, Mem. Inst. Ital. Idrobiol. 29 (suppl.).

Moriarty, D.J.W. 1973. The physiology of digestion of blue-green algae in the cichlid fish, *Tilapia nilotica*. J. Zool. 171: 25–39.

Nelson, J.S. 1976. Fishes of the world. John Wiley and Sons, New York. 416 pp.

Odum, W.E. 1968. The ecological significance of fine particle selection by striped mullet *Mugil cephalus*. Limnol. Oceanogr. 13: 92–98.

Odum, W.E. 1970. Utilization of the direct grazing and plant detritus food chains by the striped mullet *Mugil cephalus*. pp. 222–240. *In*: J.H. Steele (ed.) Marine Food Chains, University of California Press, Berkeley.

Payne, A.I. 1978. Gut pH and digestive strategies in estuarine grey mullet (Mugilidae) and tilapia (Cichlidae). J. Fish Biol. 13: 627–630.

Pearse, A.S. 1920. The fishes of Lake Valencia, Venezuela. University of Wisconsin Studies in Science 1, Madison.

Originally published in Env. Biol. Fish. 9: 137–144

The evolutionary ecology of respiratory mode in fishes: an analysis based on the costs of breathing

Donald L. Kramer
Department of Biology, McGill University, 1205 Ave. Docteur Penfield, Montreal, Quebec H3A 1B1, Canada

Keywords: Aerial respiration, Air breathing, Aquatic surface respiration, Oxygen, Hypoxia, Respiratory partitioning, Habitat selection, Behavioral ecology

Synopsis

Fishes use unimodal water breathing, unimodal air breathing, and a wide range of bimodal combinations of water and air breathing to obtain oxygen. This essay seeks to provide a theoretical framework in which to understand this diversity of respiratory mode. Consideration of oxygen as a resource shows that it is scarce in relation to demand and that an inadequate supply limits activity, growth, reproduction, and ultimately, survival. Thus, there should be strong selective pressure to maximize the efficiency of oxygen uptake. Both water breathing and air breathing require ventilation, circulation, locomotion, and respiratory structures. Each of these components has energy costs and may also be subject to costs in time, materials, or risk of predation. Consideration of these costs reveals that dissolved oxygen concentration and distance from the surface are key environmental determinants. Predation pressure and several morphological and behavioral characteristics may also have an influence. These factors are integrated into a general theory of breathing costs which permits prediction of patterns in respiratory partitioning, habitat selection, the frequency of occurrence of respiratory modes in different habitats and correlations within habitats between behavioral and morphological characteristics and respiratory mode.

Introduction

Important as the primitive lung was for the Devonian fishes, it has, apparently, not been of sufficient value to the recent fishes to be improved upon, or even retained.

Homer W. Smith (1931)

All fishes require oxygen for sustained life and activity, but there is great diversity in the way in which this need is met. The typical respiratory mode is extraction of dissolved O_2 from the water (water breathing or aquatic respiration). But some species can meet their needs completely from atmospheric O_2 (air breathing or aerial respiration). Many species combine the use of dissolved and atmospheric O_2 (bimodal breathing). Within this latter group there is extensive variation from almost complete dependence on air breathing to complete dependence on water breathing except when dissolved O_2 is extremely limited.

Why is there such diversity in respiratory mode? Discussions of the evolution of the lung (e.g. Gans 1970a, 1970b, Packard 1974) and comparative physiology (e.g. Hughes 1963, Roberts 1975, Schmidt-Nielsen 1979) have suggested several advantages to

Thomas M. Zaret (ed.), Evolutionary ecology of neotropical freshwater fishes. ISBN 90 6193 823 6

air breathing. It prevents suffocation in low concentrations of dissolved O_2 (aquatic hypoxia). It permits aerobic respiration to continue in the absence of water during periods of habitat desiccation, low tide, or voluntary emergence, and it offers an energetic advantage because of the superior properties of air as a respiratory medium. None of these hypotheses explains the success of numerous water-breathing species in hypoxic habitats (Saxena 1963, Kramer et al. 1978) and the retention and use of water-breathing capability in so many bimodal species (Johansen 1970). Such anomalies could be consequences of the dependence of evolutionary processes on preadaptations, developmental constraints, or rare random events that prevent the achievement of an 'optimal' solution to the problem of O_2 uptake. However, the frequent evolutionary loss of air breathing (Smith 1931, Gans 1970a), its variability in some closely related species (Graham et al. 1978), and the lack of consistent air breathing in bimodal organisms suggest that there are advantages to water breathing or costs to air breathing which have been overlooked. Gans (1970a, 1970b) clearly recognized this problem and proposed several factors, including problems with buoyancy and CO_2 elimination. But his discussion focussed on the evolution of gas exchange mechanisms and did not offer a general explanation for the diversity of respiratory modes in fishes.

The present essay analyzes the costs of breathing from the perspective of behavioral ecology (e.g. Krebs & Davies 1981). It suggests that habitats and species characteristics will influence the relative costs of using different O_2 sources, leading to predictions of ecological, morphological, and behavioral correlates of respiratory mode.

Oxygen as a resource

Demand for oxygen

It is usually assumed that O_2 is an essential resource for fishes; most species die within minutes without it (Doudoroff & Shumway 1970). But this is not universally the case. There are reports of fish surviving from hours to months in anoxic or near anoxic conditions in the deep waters of African lakes (Coulter 1967), in ice-covered temperate ponds (Blažka 1958), or in sealed containers (Mathur 1967, Congleton 1974). Caution is necessary in interpreting these reports because of the ability of some fish to take up dissolved O_2 even at very low concentrations (Van den Thillart & Kesbeke 1978). However, laboratory investigations have confirmed that through metabolic rate reduction, tolerance of lactate accumulation, metabolic breakdown of lactate, and the production of nontoxic anaerobic end products, some fishes can sustain life without O_2 for several days and perhaps much longer (Blažka 1958, Shoubridge & Hochachka 1981, Hochachka 1982). A mechanism for 'endogenous respiration' involving carotenoid pigments has also been proposed for fish embryos (Balon 1977, p. 172). Such tactics may also be involved in tolerance of hypoxia, even in species that cannot survive anoxia. However, the limitation on the use of these processes appears to be the low energy yield of the anaerobic pathways, about an order of magnitude less than the aerobic pathways (Hochachka & Somero 1976, Hochachka 1982). The best studied species, the crucian carp, *Carassius carassius*, and goldfish, *C. auratus*, can tolerate anoxia only at temperatures below 5° C. Studies of hibernating turtles which can tolerate anoxia for over 130 days (Ultsch & Jackson 1982) show the potential for anaerobiosis in vertebrates. Nevertheless, active life in most fishes appears to be dependent upon a continual supply of O_2.

The amount of O_2 that is required has been examined in numerous respirometric studies. Brett & Groves (1979) review suggests that a temperate fish in the 10–100 g size range at 15° C requires about 90 mg O_2 kg^{-1} h^{-1} while a tropical fish of similar size at 26° C requires about 150 mg O_2 kg^{-1} h^{-1} to meet standard metabolic requirements (minimal demand with no activity or food assimilation costs). Moderate activity such as steady swimming at one body length per second (bl s^{-1}) approximately doubles the O_2 demand. Maximal sustainable swimming speeds (2.5–6.5 bl s^{-1}) increase O_2 demand by about 10 times in active, streamlined species (Brett 1965). Consumption of a maintenance ration doubles the standard O_2 requirements, while heavy feeding increases O_2 demand by four times. Therefore, three times the standard rate may be a reasonable average value for the rate at which O_2 must be extracted from the environment. However, temperature, body size, and interspecific differences, in addition to food intake and activity will all affect this figure. Failure to meet this demand will reduce the energy available to sustain activity, growth and reproduction (Doudoroff & Shumway 1970, Fry 1971).

Availability of oxygen

The physiological uptake of oxygen depends upon the partial pressure (pO_2) differential, but the rate at which pO_2 drops as O_2 is removed from a given volume of medium depends on initial concentration ($[O_2]$). These parameters are both components of O_2 availability, therefore, and are related by standard formulas (e.g. Davis 1975).

Fishes may obtain O_2 from the atmosphere or from solution. In the atmosphere O_2 comprises a very stable 20.9% by volume or about 23.2% by mass. In solution $[O_2]$ is both lower and more variable. The $[O_2]$ at saturation, when the pO_2 of water is equal to that of the atmosphere, varies with temperature and salinity. As the temperature rises from 5° C to 30° C the saturation $[O_2]$ decreases from 12.8 mg l^{-1} to 7.6 mg l^{-1}. As the salinity rises from 0 g kg^{-1} to 35 g kg^{-1} at 25° C the saturation $[O_2]$ falls from 8.4 mg l^{-1} to 6.6 mg l^{-1} (Davis 1975). Diffusion in water is too slow to significantly affect $[O_2]$ more than a few millimeters from the air-water interface, but various mixing processes bring more water into contact with the surface. $[O_2]$ therefore depends upon the relative strength of mixing processes on the one hand and the discrepancy between photosynthesis and respiration on the other. Stagnant pools in tropical dry season forest streams, swamps shaded by forest or emergent aquatic vegetation, lake margins covered by floating vegetation and the bottom waters of lakes and ponds all tend to combine low light with reduced mixing which often lowers $[O_2]$ to 0–2 mg l^{-1} for days or months at a time (Carter & Beadle 1931, Carter 1934, Talling 1969, Dehadrai & Tripathi 1976, Ultsch 1976, Kramer et al. 1978). Waters with large standing crops of algae or submerged macrophytes frequently fluctuate on a diurnal basis between predawn minima of less than 1 mg l^{-1} and midday supersaturation exceeding 16 mg l^{-1} (e.g. Kramer et al. 1978). In the surface waters of the sea and large lakes mixing usually predominates over biological processes, and $[O_2]$ remains near saturation (normoxic). This is also generally true of rivers and streams (Pennak 1971), but slowly flowing, small, turbid, and heavily shaded streams can drop below 2 mg l^{-1} (Kramer & Graham 1976).

There is considerable microhabitat variation in $[O_2]$. Observations on clear, fast-flowing streams in Panama revealed backwaters with $[O_2]$ gradients from 7.0 to 0.7 mg l^{-1} over a distance of 1–2 m (Kramer unpublished observations). Small intermittent tributaries to these streams, an important habitat for several fish species, showed strong interpool variation with a range of $[O_2]$ from 4.7 to 0.8 mg l^{-1}. Of critical importance under some circumstances is the thin layer at the surface where diffusion can bring the $[O_2]$ close to saturation in otherwise hypoxic habitats. Lewis (1970) showed theoretically that the top mm or so could remain near saturation, even with heavy O_2 demand, and demonstrated that this was of considerable importance to fishes. In a laboratory study using nitrogen to deoxygenate the water column, Burggren (1982) measured an oxygenated layer less than 0.5 mm thick. Apparently, no field measurements have been made of actual gradients on this scale.

The relative scarcity of dissolved O_2 is emphasized by considering the weight of water containing the hourly O_2 requirement. This represents the theoretical minimal ventilation requirement if 100% of the O_2 were extracted from the water. Using the estimated need for three times the standard requirement or 450 mg O_2 kg^{-1} h^{-1}, a tropical fish breathing saturated water at 26° C (8.2 mg l^{-1}) requires the O_2 contained in 55 times its body weight each hour. At 1 mg l^{-1} its needs are met by the O_2 in 450 times its weight and at 0.45 mg l^{-1} by 1000 times its weight. Since utilization is often much less than 100%, actual ventilation will be considerably greater than the indicated values. This contrasts strongly with air breathing in which the same hourly requirements can be met by the O_2 in less than 0.2% of the body weight, but in which availability is restricted to the surface.

Mechanisms of oxygen acquisition

The relative scarcity of dissolved O_2 and the spatial limitation on atmospheric O_2 availability suggest that breathing may be a costly process for fishes. In order to understand these costs, it is necessary to review the breathing mechanisms, especially behavioral aspects which have received little previous attention.

Water breathing

The acquisition of dissolved O_2 is fundamentally similar in unimodal and bimodal species (see reviews by Hughes & Shelton 1962, Doudoroff & Shumway 1970, Randall 1970, Shelton 1970, Davis 1975, Holeton 1980, Johansen & Burggren 1980, Randall et al. 1981). At the gill O_2 diffuses across a thin epithelium while a pO_2 gradient is maintained by more or less continuous ventilation of water on one side and circulation of blood on the other. Countercurrent flows permit blood pO_2 to rise above exhalent water pO_2. Ventilation is generally achieved by a buccal and/or opercular pumping mechanism, but active species swimming above a threshold speed may use 'ram ventilation' in which forward locomotion provides the force to move water over the gills. O_2 uptake is facilitated by a large gill surface area, thin gill epithelium, high ventilation rate, and high hemoglobin concentration; response to hypoxia usually involve mechanisms to improve these characteristics.

The skin is generally a secondary site of dissolved O_2 uptake in fishes. It has a small surface area, thicker epithelium, lower blood supply and usually no opportunity for countercurrent flow, and its importance consequently decreases in hypoxic water (e.g. Berg & Steen 1965, Kirsch & Nonnotte 1977). However, larval fishes may be an exception, with a relatively larger surface to volume ratio, thinner epithelium, and even countercurrent flow (Liem 1981).

When subjected to hypoxia fish with access to the surface often rise and begin to breathe the surface layer (Lewis 1970). This aquatic surface respiration (ASR) is nearly universal among water breathing species from potentially hypoxic environments (Gee et al. 1978, Kramer & McGlure 1982, Kramer 1983). At moderate levels of hypoxia fish continue activities away from the surface, but as $[O_2]$ approaches zero, time at the surface approaches 100% (Kramer & Mehegan 1981, Kramer & McClure 1982). The median pO_2 at which fish spent 50% of their time in ASR was 14 torr in 22 temperate species ($T = 16.5°C$) and in 27 tropical species ($T = 27.5°C$) (Gee et al. 1978, Kramer & McClure 1982).

ASR can permit survival at otherwise lethal $[O_2]$ (Lewis 1970, Kramer & Mehegan 1981, Kramer & McClure 1982); it can increase blood pO_2 in goldfish (Burggren 1982) and improve growth of juvenile guppies *Poecilia reticulata* under hypoxic conditions (Weber & Kramer unpublished observations). Morphological features such as flattened heads, upturned mouths, small size and neutral or positive buoyancy probably enhance use of the surface layer (Lewis 1970). Some characoid fishes (e.g. *Colossoma* or *Piaractus*, and *Astyanax*) exhibit a facultative extension of the lower jaw in hypoxia (Branson & Hake 1972, Kramer unpublished observations). This may facilitate ASR. Thus, unimodal water breathing is not incompatible with tolerance of extreme hypoxia.

Air breathing

The acquisition of atmospheric O_2 shows important differences from water breathing. Unimodal air breathing in fishes is rare; most species take up at least a small amount of O_2 across skin or gill surfaces under normoxic conditions, even though they may rely on atmospheric O_2 alone when in air or in hypoxic water. Thus, most of our knowledge of air breathing comes from examination of bimodal species. Various aspects of air breathing have been reviewed by Carter (1957), Gans (1970a, 1970b), Johansen (1970), Roberts (1975), Graham (1976), Rahn & Howell (1976), Singh (1976), Liem (1980), and Randall et al. (1981).

Unlike the gills which appear to be evolutionarily

homologous and basically similar among species, air-breathing organs show remarkable diversity in their morphology and origins. In general, air-breathing organs sequester a bubble of air out of contact with the water but in contact with a thin epithelium through which O_2 diffuses into the blood. In contrast to water breathing, ventilation of the air breathing organ is periodic. The pO_2 at the start of each cycle depends on the percent exchange of the gas in the air-breathing organ which may be 100% or much less. Mechanisms for ventilation show as much diversity as the air-breathing organs, but they usually involve a buccal force pump. As with water breathing, O_2 uptake should be enhanced by increased surface area of the air-breathing organ, thinner epithelium and increased ventilation frequency. Air breathing also has behavioral components, swimming to and from the surface and sometimes whole body involvement in the ventilation movement (e.g. Gradwell 1971, Kramer & McClure 1980). A surprising degree of complexity in this behavior is suggested by depth dependent variation in swimming speed (Kramer & McClure 1981) and by social synchronization of air breathing among individuals (Kramer & Graham 1976).

The skin appears to be an important site for O_2 uptake in amphibious air breathers (Berg & Steen 1965, Jordan 1976, Graham 1976). It is not readily available as a site of atmospheric O_2 uptake for aquatic air breathers, but one case has been described in which a neotropical eleotrid, *Dormitator latifrons*, extends a vascular patch on top of its head into the air while floating at the surface (Todd 1973). Functionally, the absence of an enclosed air-breathing organ seems to tie *Dormitator* closely to the surface unlike other air-breathing fish which can spend considerable time away from the surface between breaths.

An alternative air-breathing mechanism is the use of a bubble in the buccal cavity to increase the pO_2 of inspired water which is then passed over the gills. This has not been studied in detail but has been suggested by observations of fishes which retain air bubbles while apparently performing ASR. It has been observed in goldfish (Burggren 1982), some other cyprinids (Kramer & McClure 1982), certain neotropical catfishes (Kramer 1983) and Australian gobies (Gee personal communication). If it is demonstrated that this bubble-holding actually affects O_2 uptake, such species would form an intermediate between air breathing and water breathing, obtaining O_2 directly from the atmosphere while using the mechanisms of ASR. In the present analysis, however, these species will be considered as water breathers.

Bimodal breathing

The uptake of O_2 from both dissolved and atmospheric sources involves three differences from the unimodal processes: the possession of respiratory organs appropriate to both media, the possibility of temporal variation in the relative use of the two media, and interference between the functioning of the two modes. Bimodal breathers show a great deal of variation in the development of gills and air-breathing organs (Carter 1957, Johansen 1970). Often species with well developed air-breathing organs have reduced gills, hence are 'obligate air breathers' unable to meet their O_2 demands by exclusive water breathing even in normoxic water. Other bimodal species have well developed gills and can meet their O_2 demands by water breathing down to quite low levels of dissolved $[O_2]$. Such 'facultative air breathers' may have well developed or very simple air-breathing organs.

Respiratory partitioning between atmospheric and dissolved O_2 also shows great variability among bimodal breathers (Johansen 1970, Rahn & Howell 1976, Singh 1976). Variation among species reflects the relative development of the respiratory organs as might be expected. But individuals do alter partitioning in relation to pO_2, pCO_2, temperature, level of activity, and other factors. Certain species, generally referred to as 'continuous air breathers', use some air breathing over the whole range of pO_2 while others, which may be called 'threshold air breathers', use only water except in hypoxia. Perhaps unexpectedly, the continuous air breathers include both facultative and obligate species.

Interference between respiratory modes may occur in at least two ways. For a bimodal breather using air breathing as the primary mode in hypoxic water, gills are a liability, for the very features that make them effective for water breathing permit diffusion of O_2 from the blood into the water (e.g. the gar *Lepisosteus*, Smatresk & Cameron 1982). Thicker gill epithelium and a smaller surface area may reduce this loss, but they also reduce the effectiveness of water breathing. Furthermore, some potential for O_2 loss may remain even when the respiratory surface is minized because the gills also function in CO_2 release, osmotic balance, and the excretion of nitrogenous wastes. When a bimodal fish is water breathing, the air-breathing organ may create problems because some uptake of O_2 occurs in the air-breathing organ. Since CO_2 is released at the gill surface, negative buoyancy results unless the air-breathing organ is repeatedly topped up (Gee & Graham 1978, Gee 1981). However, not all interactions between the two modes are negative. For example, Graham (personal communication) showed that acclimation to hypoxia in the loricariid catfish *Ancistrus* improved both aquatic and aerial O_2 uptake.

Costs of oxygen acquisition: a theory

The nature of the costs

Breathing may involve costs in energy, time, materials, and risk of predation. The first three are costs when their allocation to O_2 acquisition makes them unavailable to alternative fitness-related functions such as growth and reproduction. The fourth is a cost when breathing increases the risk of capture by predators. The temporal cost of breathing may also be important because it disrupts the sequences of other activities (e.g. mating in amphibians, Halliday & Sweatman 1976; foraging of amphibious mammals, Dunstone & O'Connor 1979). Each of the major functional components of O_2 acquisition, i.e. ventilation, locomotion, circulation, and the respiratory structures, may have different types of cost. The energy cost of ventilation has been the only cost considered in most previous discussions of the cost of breathing (see Jones & Schwarzfeld 1974, Holeton 1980). However, ventilation at the surface also has costs in time and risk of predation. Locomotion at the surface during ASR or to the surface for air breathing has potential costs in energy, time, and risk of predation. Since respiratory gas transport is a primary function of the circulatory system (Johansen & Burggren 1980), an important fraction of the energy cost of the cardiac pump should be considered a cost of O_2 acquisition (Jones 1971). Finally, structural costs, the energy and materials required for the development and maintenance of the gills or air-breathing organ, buccal and opercular pumps, heart and blood, must be included.

Costs of water breathing

The hypothesized, non-structural costs of water breathing are summarized in Figure 1. The energy cost of aquatic ventilation is expected to be high, even in saturated water, since water amounting to more than 50 times the body weight must be passed over the gills every hour in order to meet average O_2 demand. Efforts to estimate this cost involved some major unverified assumptions, but different approaches using different assumptions have reached similar conclusions. Critical reviews by Jones & Schwarzfeld (1974) and Holeton (1980) suggest that about 10% of the metabolic rate goes to support ventilation for fish at rest in saturated water. In general, the cost is unlikely to be less than 10% of the metabolic rate at higher levels of activity. However, fish that use ram ventilation may be an exception, since transferring the work of breathing from the muscles of the head to muscles of the body and improved swimming hydrodynamics resulting

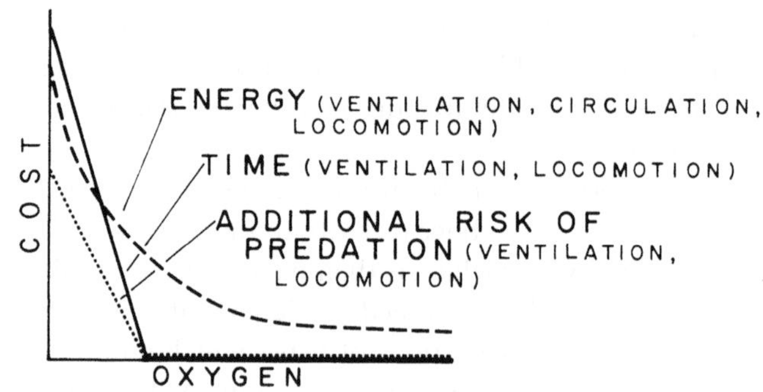

Fig. 1. The hypothesized effects of oxygen concentration on costs of water breathing for a fish capable of aquatic surface respiration (ASR). Costs of time and additional risk remain at zero except during ASR.

from the continuous water flow through the opercular openings apparently result in significant energy savings (Freadman 1981). Brown & Muir (1970) calculated the cost of ram ventilation as 1–3% of metabolic rate in the tuna. For two marine species Freadman's (1981) analysis suggests about 8% as the cost of active ventilation before the transition to ram ventilation.

With declining [O_2] energy costs should increase exponentially because the total volume ventilated will increase faster than the decline in [O_2] due to reduced percent utilization under conditions of low pO_2 and high flow, and because the cost per unit volume will increase with the total volume ventilated (Hughes & Shelton 1962, Shelton 1970, Jones 1971). Evidence for this increased cost is limited. Fish subjected to declining [O_2] in respirometers sometimes increase their O_2 demand by up to 100% before they begin to show respiratory dependence (Fry 1971). Growth rates generally show an approximately linear reduction with declining [O_2] below a threshold near 4–5 mg l^{-1} (Brett 1979).

ASR should reduce the energy cost of ventilation in a hypoxic water column because of the higher [O_2] of the inspired water. This depends on the fish's ability to use the extremely thin, well-oxygenated layer. On the other hand, ASR incurs large temporal costs for fish that do not normally live at the air-water interface, because time spent in ASR is unavailable for most other activities. The risk of aerial predation during ASR may also be higher. The survival time of six species of water-breathing fish, preyed on by a heron in the laboratory, was reduced when lower [O_2] induced higher rates of ASR (Kramer et al. 1983).

Locomotion increases the costs of water breathing only when ASR is used. At very low [O_2] this effect is due only to horizontal movements because most of the time is spent at the surface, but under less extreme conditions, the energy costs of vertical movements may reduce the energy saved in ventilation.

An analysis by Jones (1971) suggests that costs of circulation are of similar magnitude to those of ventilation at normoxia; but it is not clear whether they rise in hypoxia. In trout a decreased heart rate offsets an increased stroke volume, maintaining about the same level of blood flow and, presumably, cardiac costs (Randall 1970).

Structural costs might be estimated by the proportion that the respiratory structures (gills, heart, blood, and buccal/opercular pumps) occupy of the whole body mass. Larger gills probably entail greater structural costs but increase the percent utilization and hence decrease the ventilation costs.

Costs of air breathing

Air breathing costs are summarized in Figure 2. The energy cost of ventilation should be low in air breathers as compared to water breathers because air has only about 0.001 times the density and 0.02 times the viscosity of water, as well as having a much higher [O_2]. This is one of the classical contrasts of comparative physiology (e.g. Hughes 1963, Dejours 1976, Schmidt-Nielsen 1979). A comparison is often made between water-breathing fishes and man, whose costs are on the order of 1% of total metabolism (see Hughes & Shelton 1962, Jones & Schwarzfeld 1974). This comparison is flawed by differences in total demand between homeotherms and poikilotherms. Surprisingly, there have been no studies of the costs of air ventilation in fishes. However, estimates of ventilation cost of air breathing in frogs and turtles are similar to those of water-breathing fishes (West & Jones 1975, Kinney & White 1977). Air breathing may be less effective with respect to the properties of air than gill breathing is with respect to water. This may result from the narrow ducts and limited surface areas of the air-breathing organs of many fishes, as well as the limited ventilation potential of the buccal pump (Gans 1970b). Nevertheless, energy expenditure must be reduced by a pumping frequency one or two orders of magnitude lower than that of water breathing. The energy cost of air ventilation should be no greater than that of aquatic ventilation in normoxic water and is probably somewhat less, but it may be considerably greater than the cost of ram ventilation. On the other hand, air ventilation should be much less costly than aquatic ventilation in hypoxic water.

Locomotion is an integral part of air breathing in any fish which does not remain at the surface. A

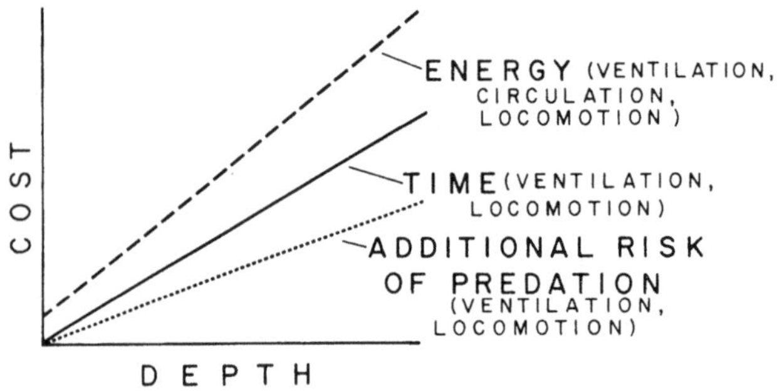

Fig. 2. The hypothesized effects of depth on costs of air breathing. At the surface, only the energy costs of ventilation and circulation are significant.

10 cm fish living 2 m from the surface and breathing 10 times per hour travels at least 40 m (or 400 bl) per hour just for breathing. At 1 bl s^{-1} this movement requires 6.7 min per hour or 11% of the fish's time resulting in an 11% increase in energy expenditure, assuming that travel at 1 bl s^{-1} doubles metabolic rate (Brett 1965). At greater depths and breathing frequencies or swimming speeds above the optimum, the locomotion cost of air breathing would exceed the estimated ventilation cost of water breathing in normoxia. Locomotion costs may be increased by having lower buoyancy while surfacing than while diving (Gee & Graham 1978, Gee 1981) and by maintaining buoyancy at depth through intake of such large gas volumes that the fish must dive against the resistance of positive buoyancy. Using data on actual swimming speeds during breathing at different depths, Kramer & McClure (1981) concluded that in the bimodal catfish *Corydoras aeneus* locomotion costs of air breathing would surpass 10% of the metabolic rate whenever locomotion exceeded 40 bl h^{-1}. Pandian & Vivekanandan (1976) and Arunachalam et al. (1976) showed that with increasing water depth in aquaria, bimodal fish increased the daily distance travelled and decreased growth efficiency. Particularly high locomotion costs may occur in air-breathing African lungfish *Protopterus aethiopicus* and catfish *Clarias mossambicus* that live at depths in excess of 50 m and 80 m, respectively, in Lake Victoria (Bergstrand & Cordone 1971, Kudhongania & Cordone 1974).

Locomotion for breathing also has costs in time. Calculations of time budgets from studies of *Corydoras* by Kramer & McClure (1980, 1981) show that when fish were at depths of 20 cm, actual breathing time ranged between 0.03 and 1.7% of total time. Breathing at the same frequencies from a depth of 1 m would have required 0.4–23.8% of total time. The upper limits approximate the temporal cost of unimodal air breathing.

Finally, locomotion may increase the risk of predation. Even very rapid breathing trips place air-breathing fish at substantial risk from an aerial predator, and this risk increases with proximity to the surface (Kramer et al. 1983). Although some fish use the surface to escape from aquatic predators, for many species leaving cover to breathe air may also substantially increase the risk of aquatic predation. There may be trade-offs between energy, time, and risk since high swimming speeds may reduce time and risk at the expense of increased energy cost.

Circulatory costs and structural costs of air breathing are presumably similar to those of water breathing. As with water breathing one expects an inverse relationship between the structural and ventilation costs involved with the elaboration of the air-breathing organs.

Costs of bimodal breathing

Costs of bimodal breathing are summarized in Figure 3. Bimodality should reduce the cost of breathing by permitting water breathing when air breathing is very costly (e.g. greater depths) and permitting air breathing when water breathing is very costly (e.g. extreme hypoxia). The costs of this increased flexibility appear to be threefold. Development and maintenance of respiratory organs should be more costly for a bimodal than for a unimodal fish, due to the larger number of structures involved. Circulation costs may be increased through increased resistance to blood flow in gills and air-breathing organ and increased circulatory demand (Gans 1970a, 1970b). The ventilation and locomotion costs of bimodal breathing may be higher than the less costly unimodal pattern because of interference between the respiratory mo-

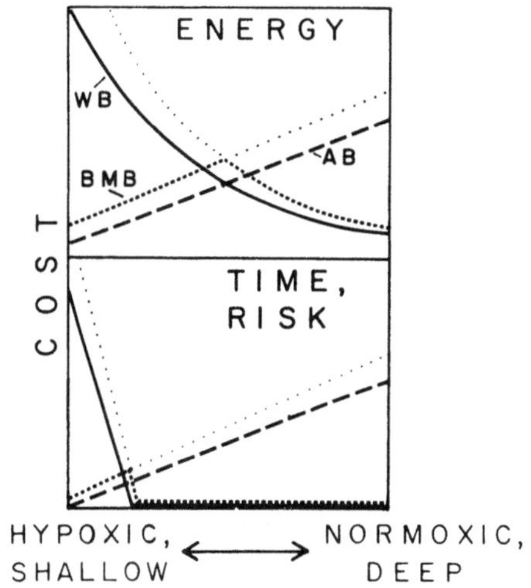

Fig. 3. The hypothesized costs of bimodal breathing (BMB) in relation to an environmental gradient from hypoxic and shallow to normoxic and deep. For comparison, costs of unimodal water breathing (WB, solid line) and unimodal air breathing (AB, long dashes) are graphed. The dotted line illustrates the costs for a bimodal breather (BMB) using exclusively water breathing or air breathing, the short dashed line indicates a bimodal breather switching between respiratory modes. Costs for using both modes simultaneously would fall between the dotted lines and short dashed lines.

des, i.e. O_2 loss from the gills when air breathing and buoyancy fluctuations when water breathing. Reduced surface areas and increased epithelial thickness of gills and air-breathing organs should lower the structural and circulatory costs and reduce interference between modes, but would also reduce the effectiveness of these organs during use.

Hence, it seems inevitable that in hypoxic water near the surface, the cost of breathing for a bimodal species would be slightly higher than that of an equivalent unimodal air breather but much lower than that of a unimodal water breather. The converse would hold in normoxic water far from the surface. At intermediate levels on a gradient from shallow, hypoxic to deep, normoxic, bimodal breathing may be more costly than both air breathing and water breathing (Fig. 3). Overall costs of breathing in bimodal forms would be lower than both unimodal patterns only in an environment fluctuating between the extremes of the shallow, hypoxic and deep, normoxic gradient. Figure 4 provides an overview of the relationships hypothesized between the costs of the respiratory modes with respect to energy, time, materials and risk.

Costs of oxygen acquisition: predictions

The theory of breathing costs makes predictions about variation in temporal costs and risk which may be measured directly. Energy costs may be estimated from physiological experimentation and biomechanical inference (Holeton 1980), but are perhaps more relevantly estimated by their effects on allocation of resources to growth and reproduction. Since organisms with lower breathing costs should have higher growth, reproduction, and survival, and thus should be favored by natural selection, a series of evolutionary predictions can also be made. Evolution should favor the lowering of breathing costs by individual responses to changing conditions, for example, respiratory partitioning and habitat selection, and by evolutionary modification of respiratory mode in response to environ-

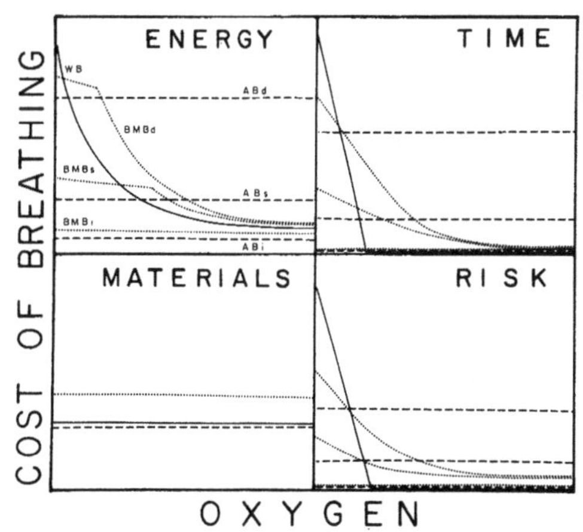

Fig. 4. Hypothesized relationships among three respiratory modes with regard to four types of breathing cost. For each cost the lines illustrate the effect of oxygen concentration on unimodal water breathing (WB, solid line), unimodal air breathing (AB, long dashes), and bimodal breathing (BMB, dots). The influence of depth on energy, time, and risk for AB and BMB is indicated by lines for fish living in deep locations (d), in shallow locations (s), and at the air-water interface (i). The BMB lines are drawn for a species which gradually increases its use of air breathing as oxygen declines.

ment and to organismal characteristics which alter breathing costs. This permits four sets of predictions which will be discussed in turn.

Respiratory partitioning

A bimodal breather can use either or both respiratory modes. If selection favors minimizing costs, variation in respiratory partitioning will be predictable. The proportional uptake of O_2 from the air should increase with hypoxia and decrease with the fish's depth. Trade-offs between the energy costs and the costs of time and risk permit predictions concerning partitioning at moderate hypoxia: increased risk of predation will increase the proportional uptake from water; activities which depend strongly upon time allocation (e.g. courtship, parental care, or feeding) will increase the proportional uptake from water; serious energy limitation will favor increased uptake from air. Finally, interactions between the water breathing and air breathing systems permit additional predictions: a species will obtain more O_2 from the air when maintaining neutral than when maintaining negative buoyancy; as total O_2 demand increases, fish will increase the proportional O_2 uptake by whichever system was initially being used less because of the exponential rise in cost with demand (Hughes & Shelton 1962, Jones 1971). All but the last two predictions also apply to the use of aquatic surface respiration versus sub-surface water breathing.

Habitat selection behavior

Natural selection should favor choosing the habitat with lowest cost. Since this prediction assumes that all else is equal, it is likely to be testable only in the laboratory. More interesting is the prediction that the strength of the preference for one habitat over another should be proportional to the difference in cost between them. When an organism chooses its habitat on the difference between its benefits and costs, the greater the cost of breathing in a habitat, the greater the benefit necessary to reverse that preference.

Distribution of respiratory modes between habitats

In a species otherwise adapted to a particular habitat type, natural selection should favor the evolution of the respiratory mode with the lowest cost. On a shorter time scale, the success of a species in colonizing a habitat should be influenced by its costs of breathing in that habitat. Thus the cost of breathing theory permits the following predictions concerning the distribution and abundance of fishes with different respiratory modes. Stable O_2 regimes will favor unimodal water breathing if normoxic and unimodal air breathing if hypoxic, while fluctuations between normoxia and hypoxia will favor bimodality. Fluctuations in depth, whether produced by environmental factors or the fish's movements, will favor bimodal breathing, while stable depth distributions will favor unimodal air breathing nearer the surface and unimodal water breathing at greater depths. Within a given depth and O_2 regime higher predation will favor water breathing over bimodal breathing, as well as increased water breathing capacity within the bimodal species. Habitats subject to hypoxia with little opportunity for ASR (e.g. burrows, floating mats of vegetation) will favor air breathing and bimodal breathing over water breathing.

Correlates of respiratory mode within habitats

In addition to predictions of differences between habitats, it should be possible to predict correlations within habitats between respiratory mode and morphological, physiological or behavioral characteristics of a species which differentially affect the cost of water and air breathing. For example, if lack of a buoyancy organ, large body size, ventral mouth position, or elongate jaw structures greatly increase the cost of ASR, fishes with these features are more likely to be capable of air breathing than to be unimodal water breathers, if they occur in habitats that ever become hypoxic. If water breathing by ram ventilation is less costly than by buccal/opercular pumping, then very active species are more likely to be unimodal water breathers than less active species, even if habitually found near the surface. If large body size, bony plates, or other characteristics reduce the risk of predation, fish with these features are more likely to be capable of air breathing than are fish which live in similar environments but lack this protection.

Alternative hypotheses

The previous section showed how the cost of breathing could predict the distribution and use of respiratory modes. But several alternative hypotheses for the evolution of breathing modes have been proposed. They are not mutually exclusive, and, as Gans (1970b) has emphasized, there is no reason to believe that a single factor should account for the diversity of respiratory modes. However, it is desirable to separate the hypotheses because each makes some different predictions.

Prevention of suffocation

Air breathing may be an adaptation to prevent suffocation when water breathing is impossible due to lack of dissolved O_2 or lack of water (Hora 1935, Carter 1957). In some physiological studies it is assumed that bimodal species breathe air only to make up for a deficit in their capacity to take up dissolved O_2. This hypothesis does not take into account that water-breathing fish without specialized air-breathing structures can survive both hypoxic water and air exposure (e.g. *Rivulus*, Cyprinodontidae, Brockmann 1975, Seghers personal communication). Nevertheless, survival of lethal conditions may be a stronger selection pressure than cost minimization.

Toxins and interfering substances

Dissolved toxic substances could favor air breathing to minimize contact between the gill epithelium and water. High pCO_2 has been suggested as such a factor (Saxena 1963); other possibilities include low pH and high $[H_2S]$. Anything which directly interferes with gill function, such as high densities of suspended clay particles, could have similar effects. As with the prevention of suffocation, such theories explain the evolution of air breathing more readily than the maintenance of water breathing or the range of bimodality.

Metabolic scope

Johansen (1970) and Packard (1974) argued that a selective advantage of air breathing was an increased potential for aerobic activity. There is an upper limit to the amount of O_2 which can be taken up by any respiratory organ (Hughes & Shelton 1962). As this limit is approached, the work expended per mg O_2 obtained increases; hence an increasing proportion of total O_2 goes to support the work of breathing. The point at which the O_2 supplied to non-respiratory tissues starts to decline with further ventilation sets the upper limit to activity (Jones 1971). If air breathing requires less work to acquire a given amount of O_2, we may expect that the maximum amount of O_2 that can be supplied to the tissues will also be higher, as Packard (1974) concluded. However, when locomotor and cardiac as well as ventilatory work are taken into account, it seems that air breathing will not necessarily provide the larger scope. This hypothesis has been supported so far by studies on the relationship between activity and use of air breathing in bimodal fishes, but not by the required comparisons between the scope of bimodal and unimodal water-breathing species.

CO_2 elimination

Gans (1970b) proposed that the ease of CO_2 elimination into water favors the retention of gills in air-breathing fish. This leads to the conclusion that bimodal fish may have gills that are disadvantageous from the viewpoint of O_2 acquisition. It offers an alternative explanation for the general lack of unimodal air breathers which must otherwise be unadaptive or due to an absence of environments that favor this specialization.

Phylogeny

If the evolution of respiratory modes is slow relative to the evolutionary divergence of taxa, then respiratory modes would be more predictable from phylogenetic relationships than from the costs of breathing. Therefore, the diversity presently observed would have a basis in historical events and selection pressures which cannot now be examined or would be due to developmental interactions rather than adaptation (Gould & Lewontin 1979). This hypothesis may be assumed by authors who have proposed advantages to air breathing without suggesting counteracting disadvantages.

Conclusions

The widespread occurrence of unimodal water breathing fishes and the extensive use of water-breathing by bimodal forms suggests to me that there must be some advantages to water breathing that counteract the well known advantages of air breathing. The key contribution of my approach has been the suggestion that water breathers avoid the high travel costs of air breathing. This has led to the prediction that distance from the surface will be as important a factor as aquatic hypoxia in the evolution and use of air breathing in aquatic forms. The cost of breathing theory is more consistent with widespread water breathing in fishes than theories which propose only unidirectional advantages to air breathing. It also has the heuristic value of incorporating respiratory partitioning, habitat selection, ecological distributions, and the evolution of breathing modes into a single framework.

The theory was developed with reference to fishes, especially tropical freshwater forms, and may owe more than I recognize to my knowledge of these organisms. However, it should be broadly applicable, not only to fishes in other environments, but to other taxa capable of air and water breathing, including amphibians, aquatic insects and gastropod molluscs.

Acknowledgements

My research was supported by a postdoctoral fellowship from the Smithsonian Tropical Research Institute and by grants from the Natural Science and Engineering Research Council of Canada and from McGill University. The governments of Panama, Brazil and Trinidad permitted my research. I am particularly grateful to Jeff Graham, Jack Gee and Karel Liem and my students, John Mehegan, Martha McClure, Anne Braun, Jean-Michel Weber, André Talbot, Dave Manley and Ray Bourgeois, for much useful stimulation and discussion. The Friday Harbor Laboratories of the University of Washington provided an excellent environment in which to write. An earlier draft received criticism from J. Dunn, J. Graham, S. Kerr, P. Hochachka, K. Liem, M. Power, R. Warner and T. Zaret. V. Kramer, W. Milsom and R. Strathmann discussed the ideas with me. Although they do not necessarily agree with my conclusions, their comments have been extremely helpful.

References cited

Arunachalam, S., E. Vivekanandan & T.J. Pandian. 1976. Food intake, conversion and swimming activity in the air-breathing catfish *Heteropneustes fossilis*. Hydrobiologia 51: 213–217.

Balon, E.K. 1977. Early ontogeny of *Labeotropheus* Ahl, 1927 (Mbuna, Cichlidae, Lake Malawi), with a discussion on advanced protective styles in fish reproduction and development. Env. Biol. Fish. 2: 147–176.

Berg, J. & J.B. Steen. 1965. Physiological mechanism for aerial respiration in the eel. Comp. Biochem. Physiol. 15: 469–484.

Bergstand, E. & A.J. Cordone. 1971. Exploratory bottom trawling in Lake Victoria. Afr. J. Trop. Hydrobiol. Fish. 1: 13–23.

Blažka, P. 1958. The anaerobic metabolism of fish. Physiol. Zool. 31: 117–128.

Branson, B.A. & P. Hake. 1972. Observations on an accessory breathing mechanism in *Piaractus nigripinnis* (Cope) (Pisces: Teleostomi: Characidae). Zool. Anz. 189: 292–297.

Brett, J.R. 1965. The relation of size to rate of oxygen consumption and sustained swimming speed of sockeye salmon (*Oncorhynchus nerka*). J. Fish. Res. Board Can. 22: 1491–1501.

Brett, J.R. 1979. Environmental factors and growth. pp. 599–675. *In:* W.S. Hoar, D.J. Randall & J.R. Brett (ed.) Fish Physiology, Vol. 8, Academic Press, New York.

Brett, J.R. & T.D.D. Groves. 1979. Physiological energetics. pp. 279–352. *In:* W.S. Hoar, D.J. Randall & J.R. Brett (ed.) Fish Physiology, Vol. 8, Academic Press, New York.

Brockmann, F.W. 1975. An unusual habitat for the fish *Rivulus marmoratus*. Florida Sci. 38: 35–36.

Brown, C.E. & B.S. Muir. 1970. Analysis of ram ventilation of fish gills with application to skipjack tuna (*Katsuwonus pelamis*). J. Fish. Res. Board Can. 27: 1637–1652.

Burggren, W.W. 1982. 'Air gulping' improves blood oxygen

transport during aquatic hypoxia in the goldfish, *Carassius auratus*. Physiol. Zool. 55: 327–334.

Carter, G.S. 1934. Results of the Cambridge expedition to British Guiana, 1933. The fresh waters of the rain-forest areas of British Guiana. J. Linn. Soc. Lond. (Zool.) 39: 147–193.

Carter, G.S. 1957. Air breathing. pp. 65–79. *In:* M.E. Brown (ed.) Physiology of Fishes, Vol. 1, Academic Press, London.

Carter, G.S. & L.C. Beadle. 1931. The fauna of the swamps of the Paraguayan chaco in relation to its environment – II. Respiratory adaptations in the fishes. J. Linn. Soc. Lond. (Zool.) 37: 327–368.

Coulter, G.W. 1967. Low apparent oxygen requirements of deep-water fishes in Lake Tanganyika. Nature, Lond. 215: 317–318.

Congleton, J.L. 1974. The respiratory response to asphyxia of *Typhlogobius californiensis* (Teleostei: Gobiidae) and some related gobies. Biol. Bull. 146: 186–205.

Davis, J.C. 1975. Minimal dissolved oxygen requirements of aquatic life with emphasis on Canadian species: a review. J. Fish. Res. Board Can. 32: 2295–2332.

Dehadrai, P.V. & S.D. Tripathi. 1976. Environment and ecology of freshwater air-breathing teleosts. pp. 39–72. *In*: G.M. Hughes (ed.) Respiration of Amphibious Vertebrates, Academic Press, London.

Dejours, P. 1976. Water versus air as the respiratory media. pp. 1–15. *In*: G.M. Hughes (ed.) Respiration of Amphibious Vertebrates, Academic Press, London.

Doudoroff, P. & D.L. Shumway. 1970. Dissolved oxygen requirements of freshwater fishes. Food and Agricultural Organization of the United Nations Technical Paper 86. 291 pp.

Dunstone, N. & R.J. O'Connor. 1979. Optimal foraging in an amphibious mammal. I. The aqualung effect. Anim. Behav. 27: 1182–1194.

Freadman, M.A. 1981. Swimming energetics of striped bass (*Morone saxatilis*) and bluefish (*Pomatomus saltatrix*): hydrodynamic correlates of locomotion and gill ventilation. J. Exp. Biol. 90: 253–265.

Fry, F.E.J. 1971. The effect of environmental factors on the physiology of fish. pp. 1–98. In: W.S. Hoar & D.J. Randall (ed.) Fish Physiology, Vol. 6, Academic Press, New York.

Gans, C. 1970a. Respiration in early tetrapods – the frog is a red herring. Evolution 24: 723–734.

Gans, C. 1970b. Strategy and sequence in the evolution of the external gas exchangers of ecothermal vertebrates. Forma et Functio 3: 61–104.

Gee, J.H. 1981. Coordination of respiratory and hydrostatic function of the swimbladder in the central mudminnow, *Umbra limi*. J. Exp. Biol. 92: 37–52.

Gee, J.H. & J.B. Graham. 1978. Respiratory and hydrostatic functions of the intestine of the catfishes *Hoplosternum thoracatum* and *Brochis splendens* (Callichthyidae). J. Exp. Biol. 74: 1–16.

Gee, J.H., R.F. Tallman & H.J. Smart. 1978. Reactions of some great plains fishes to progressive hypoxia. Can. J. Zool. 56: 1962–1966.

Gould, S.J. & R.C. Lewontin. 1979. The spandrels of San Marco and the panglossian paradigm: a critique of the adaptationist programme. Proc. Royal Soc. Lond., Ser. B. 205: 580–599.

Gradwell, N. 1971. A photographic analysis of the air breathing behavior of the catfish, *Plecostomus punctatus*. Can. J. Zool. 49: 1089–1094.

Graham, J.B. 1976. Respiratory adaptations of marine air-breathing fishes, pp. 165–187. *In*: G.M. Hughes (ed.) Respiratory Adaptations of Amphibious Vertebrates, Academic Press, London.

Graham, J.B., D.L. Kramer & E. Pineda. 1978. Comparative respiration of an air-breathing and a non-air-breathing characoid fish and the evolution of aerial respiration in characins. Physiol. Zool. 51: 279–288.

Halliday, T.R. & H.P.A. Sweatman. 1976. To breathe or not to breathe: the newt's problem. Anim. Behav. 24: 551–561.

Hochachka, P.W. 1982. Anaerobic metabolism: living without oxygen. pp. 138–150. *In*: C.R. Taylor, K. Johansen & L. Bolis (ed.) A Companion to Animal Physiology, Cambridge Univ. Press, Cambridge.

Hochachka, P.W. & G.N. Somero. 1976. Enzyme and metabolic adaptations to low oxygen. pp. 279–314. *In*: R.C. Newell (ed.) Adaptation to Environment: Essays on the Physiology of Marine Animals, Butterworths, London.

Holeton, G.F. 1980. Oxygen as an environmental factor of fishes. pp. 7–32. *In*: M.A. Ali (ed.) Environmental Physiology of Fishes, Plenum, New York.

Hora, S.L. 1935. Physiology, bionomics and evolution of air-breathing fishes of India. Trans. Nat. Inst. Sci. India 1: 1–16.

Hughes, G.M. 1963. Comparative physiology of vertebrate respiration. Heinemann Educational Books, London. 145 pp.

Hughes, G.M. & G. Shelton. 1962. Respiratory mechanisms and their nervous control in fish. pp. 275–364. *In*: O. Lowenstein (ed.) Advances in Comparative Physiology and Biochemistry, Vol. 1, Academic Press, London.

Johansen, K. 1970. Air breathing in fishes. pp. 361–411. *In*: W.S. Hoar & D.J. Randall (ed.) Fish Physiology, Vol. 4, Academic Press, New York.

Johansen, K. & W.W. Burggren. 1980. Cardiovascular function in the lower vertebrates. pp. 61–117. *In*: G.H. Bourne (ed.) Hearts and Heart-Like Organs, Vol. 1, Academic Press, New York.

Jones, D.R. 1971. Theoretical analysis of factors which may limit the maximum oxygen uptake of fish: the oxygen cost of the cardiac and branchial pumps. J. Theor. Biol. 32: 341–349.

Jones, D.R. & T. Schwarzfeld. 1974. The oxygen cost to the metabolism and efficiency of breathing in trout (*Salmo gairdneri*). Respir. Physiol. 21: 241–254.

Jordan, J. 1976. The influence of body weight on gas exchange in the air-breathing fish, *Clarias batrachus*. Comp. Biochem. Physiol. 53A: 305–310.

Kinney, J.L. & F.N. White. 1977. Oxidative cost of ventilation in a turtle *Pseudemys floridana*. Resp. Physiol. 31: 327–332.

Kirsch, R. & G. Nonnotte. 1977. Cutaneous respiration in three freshwater teleosts. Respir. Physiol. 29: 339–354.

Kramer, D.L. 1983. Aquatic surface respiration in the fishes of Panama: distribution in relation to risk of hypoxia. Env. Biol.

Fish. 8: 49–54.
Kramer, D.L. & J.B. Graham. 1976. Synchronous air breathing, a social component of respiration in fishes. Copeia 1976: 689–697.
Kramer, D.L., C.C. Lindsey, G.E.E. Moodie & E.D. Stevens. 1978. The fishes and the aquatic environment of the central Amazon basin, with particular reference to respiratory patterns. Can. J. Zool. 56: 717–729.
Kramer, D.L., D. Manley & R. Bourgeois. 1983. The effect of respiratory mode and oxygen concentration on the risk of aerial predation in fishes. Can. J. Zool. 61: 653–665.
Kramer, D.L. & M. McClure. 1980. Aerial respiration in the catfish, *Corydoras aeneus* (Callichthyidae). Can. J. Zool. 58: 1984–1991.
Kramer, D.L. & M. McClure. 1981. The transit cost of aerial respiration in the catfish *Corydoras aeneus* (Callichthyidae). Physiol. Zool. 54: 189–194.
Kramer, D.L. & M. McClure. 1982. Aquatic surface respiration, a widespread adaptation to hypoxia in tropical freshwater fishes. Env. Biol. Fish. 7: 47–55.
Kramer, D.L. & J.P. Mehegan. 1981. Aquatic surface respiration, an adaptive response to hypoxia in the guppy, *Poecilia reticulata* (Pisces, Poeciliidae). Env. Biol. Fish. 6: 299–313.
Krebs, J.R. & N.B. Davies. 1981. An introduction to behavioural ecology. Sinauer Associates, Sunderland. 292 pp.
Kudhongania, A.W. & A.J. Cordone. 1974. Batho-spatial distribution patterns and biomass estimate of the major demersal fishes in Lake Victoria. Afr. J. Trop. Hydrobiol. Fish. 3: 15–31.
Lewis, W.M., Jr. 1970. Morphological adaptations of cyprinodontoids for inhabiting oxygen deficient waters. Copeia 1970: 319–326.
Liem, K.F. 1980. Air ventilation in advanced teleosts: biomechanical and evolutionary aspects. pp. 57–91. *In*: M.A. Ali (ed.) Environmental Physiology of Fishes, Plenum, New York.
Liem, K.F. 1981. Larvae of air-breathing fishes as countercurrent flow devices in hypoxic environments. Science 211: 1177–1179.
Mathur, G.B. 1967. Anaerobic respiration in a cyprinoid fish *Rasbora daniconius* (Ham.). Nature, Lond. 214: 318–319.
Packard, G.C. 1974. The evolution of air-breathing in paleozoic gnathostome fishes. Evolution 28: 320–325.
Pandian, T.J. & E. Vivekanandan. 1976. Effects of feeding and starvation on growth and swimming activity in an obligatory air-breathing fish. Hydrobiologia 49: 33–39.
Pennak, R.W. 1971. Towards a classification of lotic habitats. Hydrobiologia 38: 321–334.
Rahn, H. & B.J. Howell. 1976. Bimodal gas exchange. pp. 271–285. *In*: G.M. Hughes (ed.) Respiration of Amphibious Vertebrates, Academic Press, London.
Randall, D.J. 1970. Gas exchange in fish. pp. 253–292. *In*: W.S. Hoar & D.J. Randall (ed.) Fish Physiology, Vol. 4, Academic Press, New York.
Randall, D.J., W.W. Burggren, A.P. Farrell & M.S. Haswell. 1981. The evolution of air breathing in vertebrates. Cambridge Univ. Press, Cambridge. 133 pp.
Roberts, J.L. 1975. Respiratory adaptations of aquatic animals. pp. 395–414. *In*: F.J. Vernberg (ed.) Physiological Adaptation to the Environment, Intext Educational, New York.
Saxena, D.B. 1963. A review on ecological studies and their importance in the physiology of air-breathing fishes. Ichthyologica 2: 116–128.
Schmidt-Nielsen, K. 1979. Animal physiology: adaptation and environment. 2nd ed, Cambridge Univ. Press, Cambridge. 560 pp.
Shelton, G. 1970. The regulation of breathing. pp. 293–359. *In*: W.S. Hoar & D.J. Randall (ed.) Fish Physiology, Vol. 4, Academic Press, New York.
Shoubridge, E.A. & P.W. Hochachka. 1981. The origin and significance of metabolic carbon dioxide production in the anoxic goldfish. Mol. Physiol. 1: 315–338.
Singh, B.N. 1976. Balance between aquatic and aerial respiration. pp. 125–164. *In*: G.M. Hughes (ed.) Respiration of Amphibious Vertebrates, Academic Press, London.
Smatresk, N.J. & J.N. Cameron. 1982. Respiration and acid-base physiology of the spotted gar, a bimodal breather. I. Normal values and the response to severe hypoxia. J. Exp. Biol. 96: 263–280.
Smith, H.W. 1931. Observations on the African lungfish *Protopterus aethiopicus*, and on evolution from water to land environment. Ecology 12: 164–181.
Talling, J.F. 1969. The incidence of vertical mixing, and some biological and chemical consequences in tropical African lakes. Verh. Internat. Verein. Limnol. 17: 998– 1012.
Todd, E.S. 1973. Positive buoyancy and air-breathing: a new piscine gas bladder function. Copeia 1973: 461–464.
Ultsch, G.R. 1976. Eco-physiological studies of some metabolic and respiratory adaptations of sirenid salamanders. pp. 287–312. *In*: G.M. Hughes (ed.) Respiration of Amphibious Vertebrates, Academic Press, London.
Ultsch, G.R. & D.C. Jackson. 1982. Long-term submergence at 3° C of the turtle, *Chyrsemys picta belli*, in normoxic and severely hypoxic water. I. Survival, gas exchange and acid-base status. J. Exp. Biol. 96: 11–28.
Van den Thillart, G. & F. Kesbeke. 1978. Anaerobic production of carbon dioxide and ammonia by goldfish *Carassius auratus* (L.). Comp. Biochem. Physiol. 59A: 393–400.
West, N.H. & D.R. Jones. 1975. Breathing movements in the frog *Rana pipiens*. II. The power output and efficiency of breathing. Can. J. Zool. 53: 345–353.

Originally published in Env. Biol. Fish. 9: 145–158

Evolutionary genetics of trophic differentiation in goodeid fishes of the genus *Ilyodon*

Bruce J. Turner[1], Thaddeus A. Grudzien[1], Karen P. Adkisson [2] & Matthew M. White[1]
[1] *Department of Biology, Virginia Polytechnic Institute and State University, Blacksburg, VA 24061, U.S.A.*
[2] *Department of Biology, Roanoke College, Salem, VA 24153, U.S.A.*

Keywords: Polymorphism, Allozymes, Chromosomes, Geographic variation, Disruptive selection, Morphological systematics, Atheriniformes

Synopsis

Some populations of one or more species of the goodeid fish genus *Ilyodon* in certain tributaties of Coahuayana and Armeriá rivers (Jalisco and Colima, Mexico) are dichotomous with respect to morphological features that are presumptive trophic adaptations. A narrow mouth 'morph' (described as *I. furcidens* in the Río Armería) is sympatric with a broad mouth morph (named *I. xantusi*). The morphs are additionally divergent in tooth and gill raker numbers and in coloration of mature males. Other populations are essentially continuous ('non-dichotomous') in these features.

An extensive allozyme survey of sympatric narrow and broad mouth morphs from four localities in the Río del Tule (a tributary of the Río Tuxpan in the Río Coahuayana basin) revealed striking geographic variation in gene frequencies between populations but no differences between morphs at any one locality. The data are consistent with the hypothesis that the morphs are components of the same gene pool, i.e. they are conspecific. This hypothesis receives additional support from analysis of the broods of field-impregnated females.

A chromosomal polymorphism, probably involving pericentric inversions, exists at one or more sites in the Río del Tule system. Individuals have 0–4 metacentric chromosomes. The frequency distribution of metacentric chromosome phenotypes (homologs cannot be distinguished) at this site is divergent between the morphs. The chromosomal polymorphism of the Rio del Tule *Ilyodon* appears to be part of a 'step cline' in the number of metacentrics among the *Ilyodon* of the Rio Coahuayana basin.

It is hypothesized that the *Ilyodon* morphs in the Rio del Tule are in genetic contact, but that disruptive selection acts to eliminate individuals with intermediate trophic phenotypes. At one site, at least, the selection is sufficiently potent to foster chromosomal differentiation between the morphs. The biological basis of the natural selection is unknown, but availability of food resources is implicated by circumstantial evidence. The genetic basis of the morphological differentiation, including the possibility of an ecophenotypic component to the variation, is currently under investigation.

Introduction

The evolution of a broad array of trophic adaptations (morphological features involved in feeding) is part and parcel of the explosive adaptive radiation of the actinopterygian fishes. Despite their obvious importance, however, the study of these adaptations on the population or species level is in its infancy. For example, we do not know if divergent trophic features are the result of changes in a relatively large number of genes or if they represent changes in a few key genetic elements.

Thomas M. Zaret (ed.), Evolutionary ecology of neotropical freshwater fishes. ISBN 90 6193 823 6

Resolution of this question has important consequences for our understanding of the ways in which fishes evolve: if trophic differences usually have an oligogenic basis, then ecologically significant differentiation could occur rather quickly, perhaps by processes akin to the 'transilience' models of Templeton (1980). Such mechanisms would provide obvious genetic bases for the evolution of the geologically young and trophically diverse species 'flocks' of the African rift lakes and elsewhere (Fryer & Iles 1972, Myers 1960, Greenwood 1981).

Recently, several studies have appeared which collectively suggest that trophic morphological differences do indeed have an oligogenic basis. Vrijenhoek (1978) has shown that two naturally-occurring clones of the triploid gynogen *Poeciliopsis 2 monacha-lucida* have well-marked differences in dentition and feeding behavior. Of equal importance are examples of trophic polymorphism. Examples of this phenomenon were first reported in the characoid *Saccocon* by Roberts (1974), and then suggested in the *Cichlasoma* of the Cuatro Cienegas basin of Coahuilla, Mexico by Sage & Selander (1975). In the latter case, Kornfield et al. (1982) have shown that three trophically differentiated forms (an algaldetritivore, piscivore and molluskivore, each with distinctive oral and pharyngeal dentition) are actually parts of the same gene pool. The trophic forms comprise a polymorphism. The existence of such polymorphisms suggests that the differences are oligogenic. If a large number of genes were involved, rather specialized systems would have had to evolve to assure segregation of the 'correct' combinations of genes.

The existence of trophic polymorphisms has additional implications. Such polymorphisms have been suggested as a basis for sympatric speciation (Maynard Smith 1966, Tauber & Tauber 1977). The discovery of trophic polymorphism tends to strengthen the potential role of sympatric speciation in the evolution of the rift lakes species flocks. Thus Greenwood (1981, p. 71), hardly an advocate of sympatric or 'intralacustrine' speciation, writes: 'If genes controlling these features [i.e. divergent trophic phenotypes of the Cuatro Cienegas cichlids] were to become linked with genes affecting male breeding coloration (or other reproductive features), then, through the effects of assortative mating, the morphs could become true species ...'.

At present, the generality and distribution of intraspecific trophic polymorphisms among teleosts is unknown. Turner & Grosse (1980) suggested that two morphologically (presumably trophically) distinct, sympatric forms of the goodeid genus *Ilyodon*, were conspecific. This paper discusses work on the evolutionary genetics of trophic differentiation in *Ilyodon*. We shall show that the trophic morphs are clearly in regular genetic contact, but that in one tributary the biological situation may be more complex than the term 'trophic polymorphism' suggests. Our data will address one other topic as well: the existence of unexpectedly large amounts of microgeographic genetic differentiation within continuously distributed fishes of a single river system. We hope to demonstrate the inadequacy of systematic inferences at the species and population level that are based only upon morphological features, no matter how elegantly analyzed.

General biology of Ilyodon

The genus *Ilyodon* is a member of the Goodeidae, a family (or subfamily, see Parenti 1981) of viviparous cyprinodontoid fishes mostly endemic to the Mesa Central (sensu Smith et al. 1975) of the Mexican plateau, where it has undergone an adaptive radiation of at least Miocene antiquity (Smith 1980). The general biology of the family is discussed by Miller & Fitzsimons (1971). The last major taxonomic revision was by Hubbs & Turner (1939). Reproductive biology, including references to now classic descriptive studies by C.L. Turner and G. Mendoza, is discussed by Grier et al. (1978) and especially Wourms (1981).

The center of abundance of the family is the large Río Lerma – Río Grande de Santiago system, including its associated lakes, and former affluents [geological history of the system is discussed by Barbour (1973) and Smith (1980)]. *Ilyodon* species do not occur in the contemporary Río Lerma system, but are confined to the Ríos Ameca (a few tributaries only), Armería, Coahuayana, Balsas, and some intervening coastal rivers. Our work has been restricted thus far to the basins of the Ríos Armería and Coahuayana. These rivers, which drain a southwestern portion of the plateau to the Pacific ocean, are faunally less diverse than the Río Lerma system. Their historical relationships to each other and the Río Lerma are not yet completely known. Their upper courses (on the plateau proper) contain members of another endemic goodeid genus, *Allodontichthys* (see Miller & Uyeno 1980), species of which are rheophilic, bottom – dwelling and darter-like in general behavior. Other fishes sometimes present include the goodeids *Xenotoca eiseni* and *X. melanosoma* (upper Río Coahuayana only, the species also occur in the Río Ameca, see Fitzsimons 1972), the *Notropis*-like cyprinid *Alganxa aphanea* (certain tributaries of the Río Armería and Coahuayana, see Barbour & Miller 1978), and the catostomid *Moxostoma* (upper Armería only). Sunfish (*Lepomis* spp.) have been introduced and are locally common. *Ilyodon* species occur in these rivers at altitudes greater than about 180 m. In the lower courses, *Ilyodon* have been collected together with more coastal fish species, including the characoid *Astyanax fasciatus*, the cichlid *Cichlasoma istlanum*, various gobies and an eleotrid (genera *Awaous*, *Sicydium* and *Gobiomorus*), a mullet (*Agonostomus*), and the poeciliids *Poeciliopsis* spp., and *Poecilia butleri*. *Tilapia* spp. have been introduced and are now found at many locations.

Ilyodon species, in contrast to most goodeids (which are typically lake dwellers) are fluvial and distinctly minnow-like in general appearance (Fig. 1) and behavior. They are usually less than 100 mm in length. In the rivers in which they occur, *Ilyodon* are ubiquitous (except where eliminated by pollution), frequently very abundant, and are a predominant component of the fish fauna throughout their range. They are encountered in both fast flowing riffles and more quiet backwaters, and seem well adapted for movement and dispersal in streams. *Ilyodon* have been taken in large rivers and in small tributaries (some of which are temporary) over substrates which range from boulders and large pebbles to silt, sand, and mud.

Neither the diet nor the reproductive biology of natural *Ilyodon* populations has been studied in detail. Preliminary stomach content analyses in our laboratory suggest that *Ilyodon* are primarily algal grazers, though aquatic insects are also taken. Reproductive activity (as evidenced by the appearance of gravid females and of young) appears to be seasonal, roughly January through March, and to commence earlier at lower elevations.

An unusual feature of the biology of *Ilyodon* is extreme variability in markings and color pattern. This variation has sexual, individual, population and developmental components, and involves general body shade (dark olivaceous grey to silvery), the intensity of reflective colors, the number and extent of dark body markings (most mature males have a clearly delineated horizontal stripe but some lack it entirely, some females have a distinct set of vertical bars on the flanks, others have no markings or a stripe), the presence and intensity of dark terminal or subterminal bands, basal spots, or broken bars on all single fins, and the presence and extent of white, yellow or orange pigments on the single fins. The variation, some of which, at least, appears to be genetic, is greater than that encountered in any other goodeid, and rivals that of guppies, *Poecilia reticulata*. Some additional variation is local and may be related to physical features of the environment. For example, in 1980 in a tributary of the Río Armería at Apulco, in fast moving, unusually murky water with a silty substrate, we collected a sample of *Ilyodon* which were completely silver. Similarly, the *Ilyodon* from locality 12 (Fig. 2), a tiny tributary of the Río Barreras, were taken in shallow, oily black water heavily polluted with pig manure; these were all uniformly dark brown to blackish in color. At present, we do not know how much of this latter variation is genetic and how much represents ecophenotypic responses to particular environments.

Taxonomy

Currently, three nominal species in the genus *Ilyodon* are recognized: *I. whitei* (Meek) from the Río Balsas basin, and *I. furcidens* (Jordan & Gilbert) and *I. xantusi* (Hubbs & Turner) from the Río Armería. *Ilyodon xantusi* and *I. whitei* were consid-

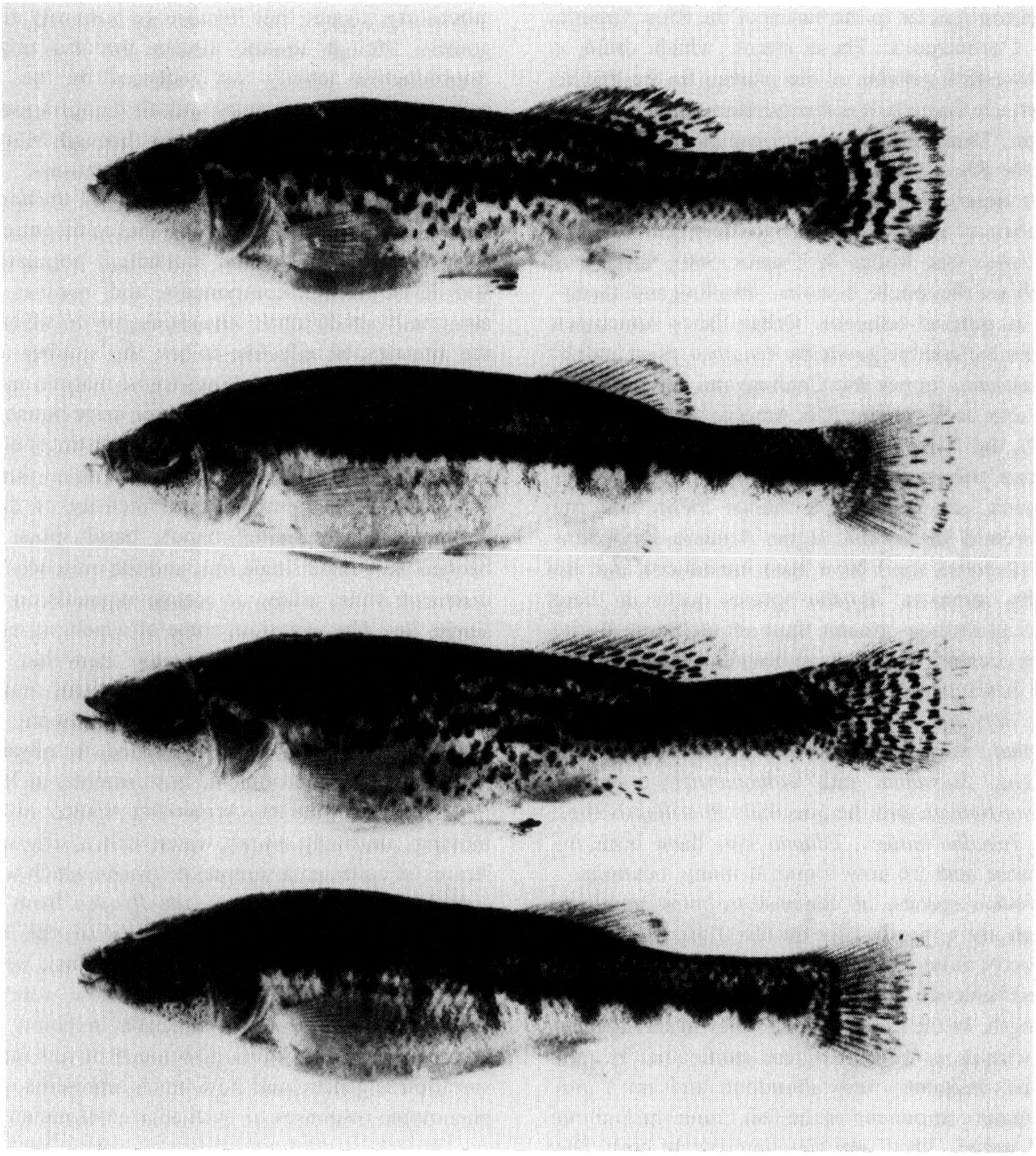

Fig. 1. Ilyodon specimens from the Río del Tule, a tributary of the Río Tuxpan system, Río Coahuayana basin. Males above, females below, in each case. Upper pair: narrow mouth morph or morph A. Lower pair: broad mouth morph or morph B. Males are 57.5 and 56 mm SL respectively. Note the terminal caudal band in the morph A male. The mouths of these specimens, which are from a dichotomous population, are shown in Figure 3A.

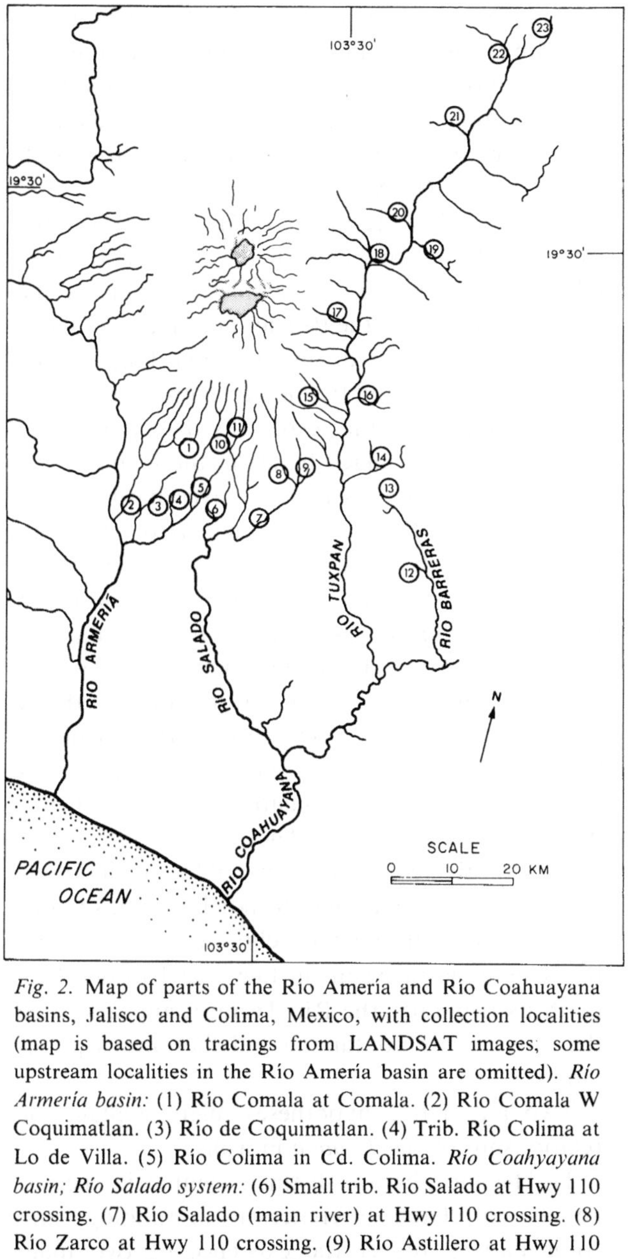

Fig. 2. Map of parts of the Río Ameria and Río Coahuayana basins, Jalisco and Colima, Mexico, with collection localities (map is based on tracings from LANDSAT images, some upstream localities in the Río Ameria basin are omitted). *Río Armeria basin:* (1) Río Comala at Comala. (2) Río Comala W Coquimatlan. (3) Río de Coquimatlan. (4) Trib. Río Colima at Lo de Villa. (5) Río Colima in Cd. Colima. *Río Coahyayana basin; Río Salado system:* (6) Small trib. Río Salado at Hwy 110 crossing. (7) Río Salado (main river) at Hwy 110 crossing. (8) Río Zarco at Hwy 110 crossing. (9) Río Astillero at Hwy 110 crossing. (10) Small trib. Río Salado at El Trapiche. (11) Small trib Río Salado at El Cobono. *Río Barreras system:* (12) Small trib ('Arroyo des Puercas') SE Pihuamo. (13) Río Barreras at Pihuamo. *Río Tuxpan system:* (14) Río del Tule and Río Terrero (see Fig. 6). (15) Río Tonila. (16) Río San Pedro (this name is also used for a trib. of the Río Ameria). (17) Río de San Marcos. (18) Río Tuxpan (main river) at Atenquique. (19) Río de San Rafael at Hwy 110 crossing. (20) Small trib. at Cd. Tamazula. (21) Río Contla at Contla. (22) Small trib. at La Garita. (23) Small trib at Puerto del Zapatero. Note: the lower course of the Río Tuxpan, below Atenquique (18), is sometimes called the 'Río Naranjo,' the upper, the 'Río Tamazula.'

ered to be generically distinct from *I. furcidens* by Hubbs & Turner (1939) who separated them in the genus *Balsadichthys*. That genus was vacated in favor of *Ilyodon* by Miller & Fitzsimons (1971). Alverez (1970, p. 98) synonymized *I. xantusi* with *I. whitei* (as *Balsadichthys* species), an action we cannot now evaluate. As used by Hubbs & Turner (1939), the terms '*furcidens*' and '*xantusi*' refer respectively to the narrow and broad mouth morphs (see below) from the Río Armería. Hubbs & Turner's species concepts were extended (with the addition of several new subspecies) by Kingston (1979) to include fishes with similar phenotypes in the Río Coahuayana basin. The work described below impeaches the current taxonomy as it applies to the dichotomous *Ilyodon* of the Río Coahuayana, but we defer formal taxonomic statements pending more data from Río Armería. In this article we restrict the use of '*xantusi*' and '*furcidens*' solely to Río Armería material, and use only the generic name in discussing Río Coahuayana *Ilyodon*. However, this does not necessarily imply that we think that the Río Coahuayana and Río Armería *Ilyodon* are specifically distinct, or that there is more than a single biological species in the genus.

The work we discuss below is based on material collected in the field in 1977, 1978, 1980 and 1981. Our collection localities in the Ríos Armería and Coahuayana are mapped in Figure 2. Some of the data will be presented in summary form only, more detailed analyses will appear elsewhere.

What is the relationship of the components of morphologically dichotomous Ilyodon *populations?*

In certain tributaries of the Río Armería and at least one tributary of the Río Coahuayana basin, the Río del Tule, adult *Ilyodon* can be separated into two groups on the basis of relative mouth width (Fig. 3). Frequency histograms of relative mouth width (mouth width/head length) or more detailed analytical treatments (Fig. 4), are clearly bimodal. These modes define two distinct morphs in what we term 'dichotomous' populations:

Morph A: narrow mouth, with essentially rounded jaws and a narrow gape, both jaws with several rows of teeth.

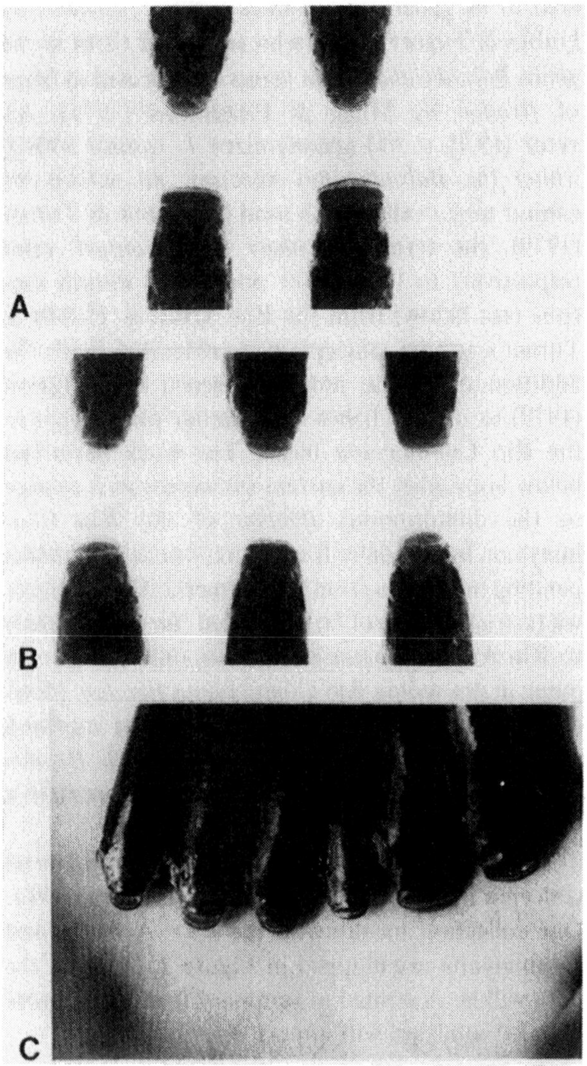

Fig. 3. Dorsal mouth profiles. (A) Dichotomous population, Río del Tule. Morph A above, morph B below (see Fig. 1). (B) Nondichotomous population, Río Tuxpan at Atenquique (locality 18). (C) Nondichotomous population, Río Colima in Cd. Colima (locality 5), hand-held field sample. Note the presence of both extreme trophic phenotypes (first and fifth specimens) and of intermediates (e.g. third specimen).

Morph B: broad mouth, with more massive, squared off jaws, a broad horizontal gape, and with inner rows of jaw teeth absent or greatly reduced. The jaw width separation is correlated with other morphological differences. These include: number of teeth in the outer row (roughly 20–30 in morph A and 40–50 in morph B in the Río del Tule), mandible length, and gill raker number (roughly 25 to 45 in morph A, 30–55 in morph B). In addition, the morphological differences are correlated with differences in coloration of sexually mature adult males. In the Río Armería, males of morph B tend to have vivid subterminal orange bands on the dorsal and anal fins and on the upper and lower edges of the caudal fin (sometimes extending to the center of the fin). In the Río del Tule, males of the A morph tend to have distinct terminal black bands on the single fins, and frequently, more vivid yellow fin coloring, single fins, especially the caudal, usually lack terminal black bands in the B morph.

In other tributaries in both river basins, most especially the Río Coahuayana, *Ilyodon* populations cannot be sorted into two groups based on the same criteria. Plots of relative mouth width in these 'nondichotomous' populations are unimodal (Fig. 4). In some nondichotomous populations, some extreme phenotypes (corresponding to the A and B morphs defined above) can be recognized, but a large proportion of each population is intermediate. In other populations extreme phenotypes are absent or rare. The male color patterns described above are also variable in the nondichotomous populations. Nondichotomous populations clearly outnumber dichotomous ones in the Río Coahuayana basin; in fact, the Río del Tule (with its tributary, the Rió Terrero) appears to be the only stream containing a dichotomous population.

Two alternative hypotheses might explain the relationship of the *Ilyodon* morphs:

1) The morphs are distinct, sympatric species, i.e. the correlated differences in morphology and color pattern reflect the existence of two independent gene pools. This hypothesis, adopted by Hubbs & Turner (1939) and Kingston (1979), follows from routine methods of classical ichthyology. It implies that the nondichotomous populations result from interspecific hybridization. This hypothesis has difficulty accounting for the existence of both dichotomous and nondichotomous populations. Why should the two species be hybridizing to a significant extent in some tributaries, but not in others?

In a Neo-Darwinian framework, this hypothesis leads to several falsifiable predictions. In tributaries where hybridization does not occur (i.e. intermediates are rare to absent), the two species should have independent gene pools. Some polymorphic gene loci might be equivalent in the two species, others might have the same secondary alleles but at statistically distinguishable frequencies, and others should be essentially fixed for alternative alleles.

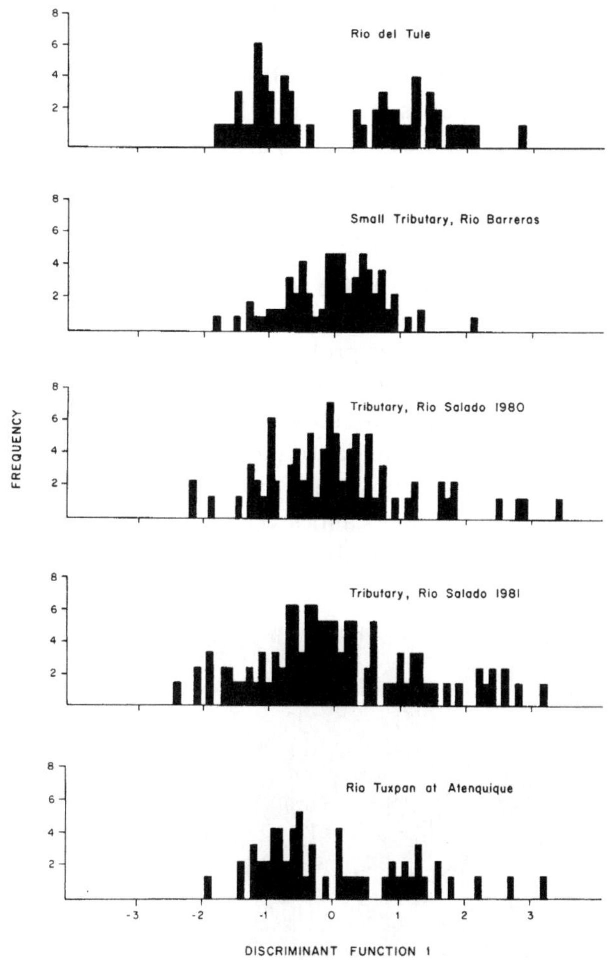

Fig. 4. Frequency histograms of discriminant function scores in four populations of *Ilyodon* from the Rio Tuxpan system. Discriminant function 1 is composed of relative mouth width and jaw length. Note that only the Río del Tule sample is dichotomous. Also note year-to-year consistency of the samples from the Río Salado (locality 6). These distributions are very similar to frequency histograms of relative mouth width alone (see Turner & Grosse 1980, Fig. 3).

Geographic variation in allele frequency, if it occurs, should be at least partially independent. The latter holds even if the variation is largely a response to natural selection; given independent evolutionary history, it is unlikely that two distinct gene pools could respond identically at more than a few loci to the same selective forces.

2) The morphs are variant components of the same gene pool, i.e. they are conspecific. The existence of both dichotomous and nondichotomous populations presents problems for this hypothesis as well. The intermediates in the nondichotomous populations are presumably the results of matings (i.e. gene flow) between the trophic morphs. If the morphs in the dichotomous populations are in genetic contact, why are there no intermediates? The presence of correlated differences in jaws, tooth number, and male color patterns also offers a problem for this hypothesis: Are all these differences the pleiotropic products of allelic differences in a single gene?

Nonetheless, the hypothesis of conspecificity also leads to some falsifiable predictions. The gene pools of the two sympatric morphs should be completely equivalent, unless acted upon by very different selective pressures. Thus, unless particular genes are directly involved in encoding the divergent morphological traits (or closely associated with such genes by genetic linkage or natural selection), allele frequencies at all polymorphic loci should be statistically indistinguishable. If geographic (or temporal) variability in allele frequencies exists, both morphs should exhibit concordant variation.

The two alternative hypotheses offer very different predictions about the gene pools of the morphs. Sympatric, conspecific morphs in regular genetic contact should probably covary. Noninterbreeding, distinct species should not. Two conditions are necessary to falsify one of these predictions:

1) Polymorphic gene loci at which allele frequencies can be measured.

The more such loci in the population, the more complete the evaluation.

2) Geographic variation in allelic frequencies.

We present genic comparisons of the morphs in the Río del Tule, a tributary of the Río Tuxpan system in the Río Coahuayana basin near Pihuamo,

Jalisco. Unless otherwise noted, all data that follow pertain to *Ilyodon* in the Río Coahuayana (the Río Armería will be considered elsewhere). The Río del Tule is a small river with a gravel and boulder substrate (Fig. 5). A small tributary, the Río Terrero, is a well known *Ilyodon* collecting and observation area (e.g. Kingston 1979). Turner & Grosse (1980) surveyed allozymes in samples of *Ilyodon* morphs taken in the Río Terrero near the town of 21 de Noviembre. There were no statistically significant differences between the morphs at five polymorphic loci. However, sample sizes were relatively small and most secondary alleles were rare. Consequently, in 1980, 4 localities in the Río del Tule system were sampled (Fig. 6) and these were surveyed for the products of 36 presumptive structural gene loci by improved electrophoretic techniques. The results, in summary form, are as follows:

1) There is very significant heterogeneity among localities; RxC/G-tests of association (Sokal & Rohlf 1981) are highly significant at 14 of a total of 15 polymorphic loci. F_{st} values (Wright 1978) for all loci are significantly greater than O (X^2 test, Eanes & Koehn 1978). Given that the entire transect is probably less than 20 km in length, the geographic variation is enormous, and is unprecedented in continuous fish populations.

2) Despite the substantial geographic variation among localities, there are no significant differences between morphs at any locality (of a total of 43 locus by locus comparisons, only 1, at La Plomoza, yielded a statistically significantly result, a level expected by chance alone).

At present we have no data that allow us to exclude either natural selection or genetic drift as a cause of the extreme microgeographic variation of allozyme frequencies detected in our *'lyodon* samples. Gradients of ecological variable. (e.g. temperature) that have been implicated as selective agents influencing some allozyme frequencies in other fish species (Johnson 1971, Mitton & Koehn 1975, Place & Powers 1979) are not evident among our collection sites. On the other hand, there are ways in which stochastic forces could affect gene frequencies. The Río Terrero has a seasonally variable flow. In the dry season, the upper courses diminish into a series of isolated or semi-isolated pools. Such drying probably results in very high mortality in the *Ilyodon* population. Severe mortality could lead to seasonal bottlenecks in the effective size of populations and thus to the loss of secondary alleles by genetic drift. Water levels at the more downstream localities, especially in the Rio del Tule itself, are probably not as affected by seasonal

Fig. 5. Río del Tule at Hwy 110 crossing 8 km N Pihuamo (station 4 in Fig. 6). Boulder-strewn substrate is typical of many *Ilyodon* habitats.

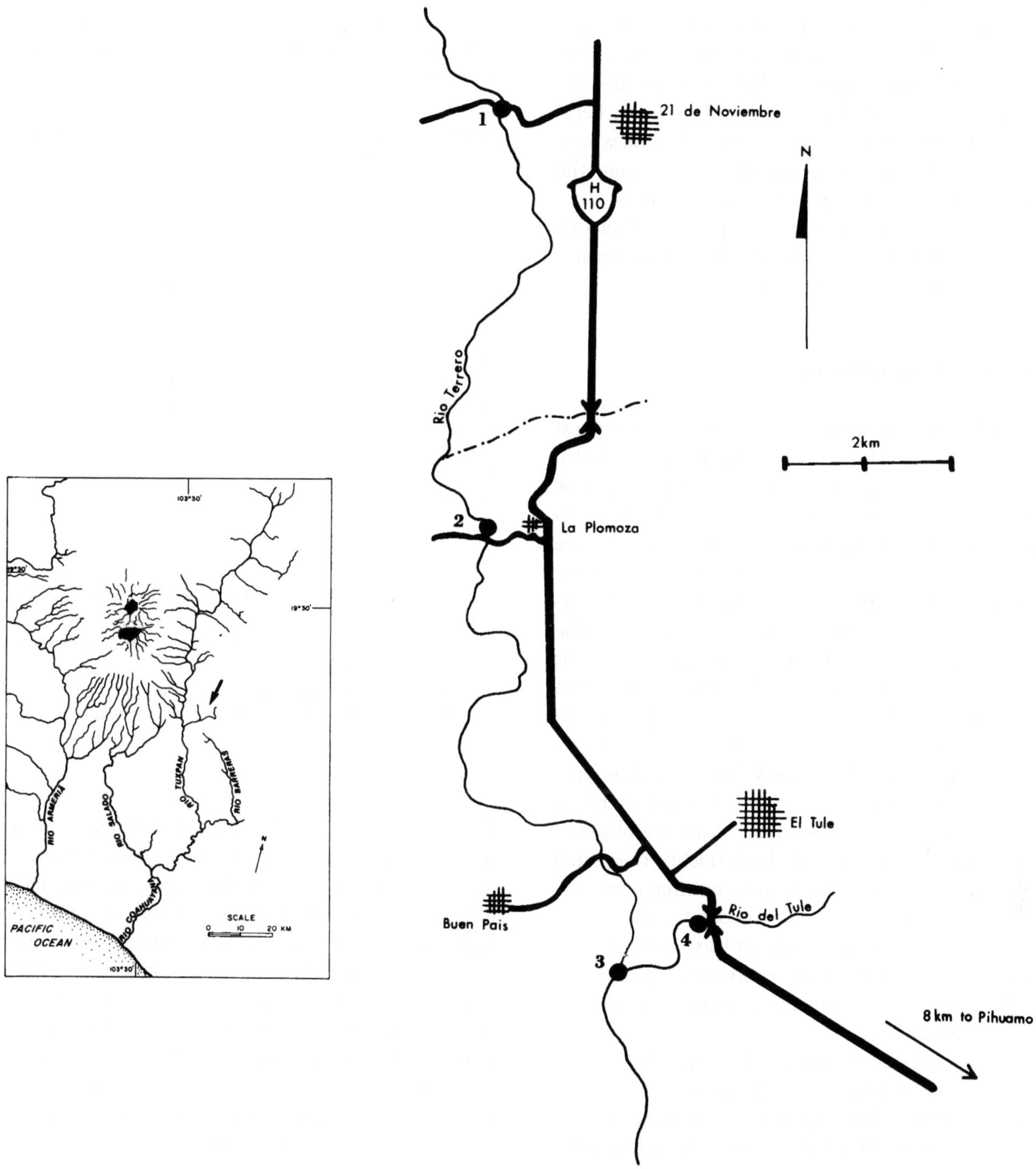

Fig. 6. Collecting stations for allozyme survey along the Río Terrero and Río del Tule (note scale).

drought, and genetic bottlenecs may be relatively infrequent. These populations would retain secondary alleles. Thus, seasonal differences in effective population size could explain the geographic heterogeneity. Regardless of the cause of the variation, the data clearly falsify the hypothesis that the morphs are distinct species. Only the most doctrinaire selectionist would attempt to explain our allozyme data by invoking absolutely parallel natural selection on two separate gene pools.

Analysis of natural broods

The allozyme data discussed above strongly suggest that there is gene flow between the *Ilyodon* morphs. However, although they indicate that gene flow exists, they provide no indication of frequency. It is axiomatic in population genetics that in the absence of selection, relatively small amounts of gene flow can prevent the divergence of geographically partially isolated populations (e.g. Spiess 1977). The same considerations hold for an ecologically or behaviorally subdivided population (providing that the genes are selectively neutral with respect to the subdivision). The allozyme data are thus compatible with levels of gene flow that range from very small (as might be the case if the morphs tend to mate assortatively) to panmixia. If the morphs tend to mate assortatively with high fidelity (mediated perhaps by either segregation into different microhabitats or behavioral cues associated with male color differences) then, under at least some circumstances, an ecological or behavioral 'species' definition might be more meaningful than a Neo-Darwinian one.

In order to assess the level of gene flow more directly, we captured gravid females in the Río Terrero in March 1981 and returned them alive to the laboratory. There they were allowed to have young, and the young were reared to a size at which morph phenotype could be scored. Data are presented in Table 1 for 14 such females (7 of each morph) and their 61 resultant progeny (mean brood size = 4.4, range 1–10). We emphasize that these data are from females taken from a natural population, not from laboratory matings. Note that 3 of

Table 1. Morph phenotype distribution among the progenies of field collected (naturally inseminated) *Ilyodon* females from the Río Terrero.

Female No.[2]	Maternal phenotype[1]	*Progeny phenotype distribution*[1] A	B	I
2	B[3]	1	2	—
3	B[3]	3	—	—
5	B[3]	1	4	2
6	A	10	—	—
7	A[3]	—	4	2
8	B[3]	3	—	—
9	B	—	1	—
12	A[3]	4	1	1
14	A[3]	—	3	—
15	A	2	—	—
16	A[3]	2	1	1
20	A	3	—	—
21	B	—	7	1
22	A	4	—	—

[1] A = narrow mouth morph, B = broad mouth morph, I = intermediate
[2] Missing numbers indicate prepartum mortality or brood mortality prior to scoring
[3] Females which produced at least one progeny with a non-maternal morph phenotype (excluding intermediates)

7 A morph females (43%) yielded at least one B morph progeny, and all 4 progeny from female 22 (an A morph) were of the B morph. We thus have direct data that relatively high levels of gene flow probably occur in natural populations. Interestingly, 7 progeny, roughly 11%, could not be readily categorized as morph A or B, and so are classified as 'I' (for intermediate). This proportion of intermediates is more characteristic of nondichotomous than of dichotomous populations. Further interpretation of this data must await the results of controlled laboratory crosses now in progress.

Chromosomal polymorphism and differentiation

A chromosomal polymorphism exists within the *Ilyodon* at at least one site, the Río Terrero near 21 de Noviembre, in the Río del Tule system. So far as we know, this polymorphism does not occur in other tributaries in the Río Coahuayana basin, though profound interpopulation differentiation is evident (see below). The polymorphism (Fig. 7, 8) is rather unusual among teleosts. It involves the number of metacentric chromosomes rather than chromosome number itself, and is therefore non-Robertsonian. Individuals may have 0 to 4 metacentrics (2N = 48). In analyzing this polymorphism, we are limited to some extent by current fish cytogenetic techniques; we cannot identify homologs with assurance. For example, among fish with 2 metacentrics, we cannot distinguish among homozygote classes (presumptive chromosome genotypes aaBB or AAbb) and a heterozygous one (chromosome genotype AaBb), and these genotypes are probably confounded in our analyses. Consequently, our interpretation may change as G-banding and other high resolution techniques for fish chromosomes are developed and applied to the problem. At present, we are restricted to simple comparisons of metacentric number classes between morphs. Such a comparison (Table 2) reveals differences among the morphs at the one locality where data are available. The differences are strikking: the modal number of metacentrics in the A morph is O, while that of the B morph is 2.

However, in view of the allozyme data and the broods from field-caught females, the chromosome data most probably do not imply that the morphs are separate gene pools. The best interpretation seems to be that the components of the chromosomal polymorphism are not selectively neutral with respect to the subdivision of the population into A and B morphs. We have not yet been able to directly evaluate this hypothesis, but we do have some evidence which suggests that there may be size correlated differences in fitness (or survival) among the chromosome phenotypes *within* the B morph. Larger individuals tend to have more metacentric chromosomes than smaller ones.

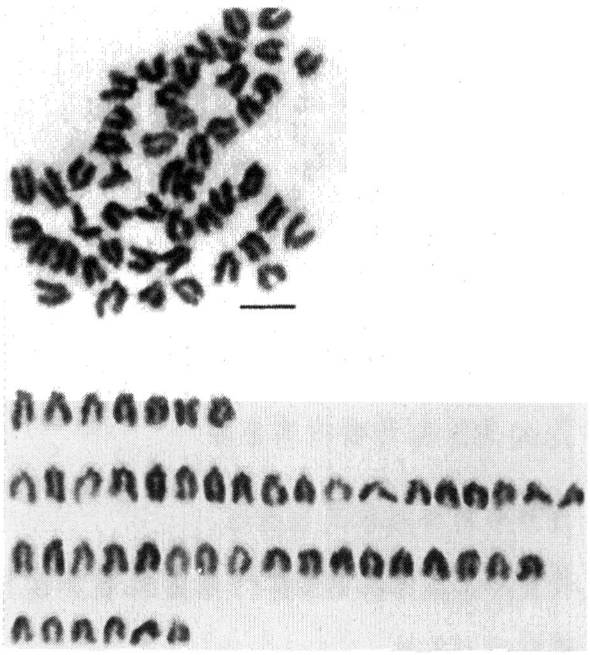

Fig. 7. Chromosomes (above) and mounted karyotype (below) of a morph A female from station 1 (Fig. 6). All chromosomes are subtelocentric to telocentric. This is an extreme phenotype of a chromosomal polymorphism (0–4 metacentrics) at this locality (see text, Table 2 and Fig. 8). All chromosome preparations are from gill epithelium, stained with Giemsa. Scale = 5 microns.

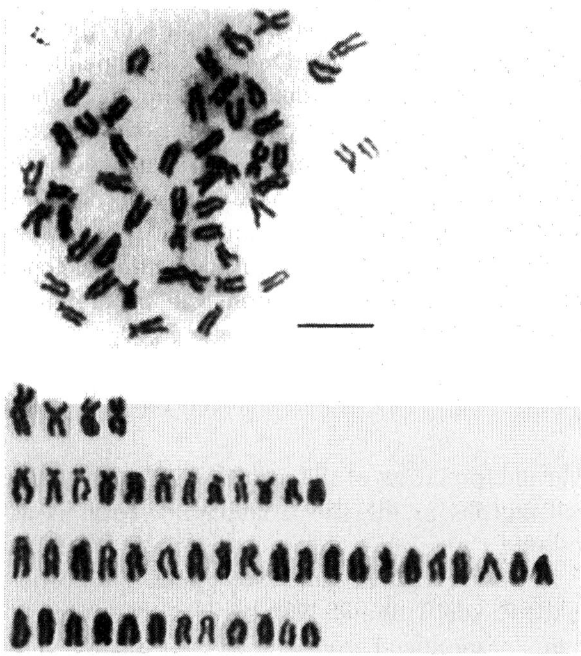

Fig. 8. Chromosomes are karyotype of a morph B male from station 1 (Fig. 6). Note 4 metacentric chromosomes, an extreme phenotype of a chromosome polymorphism at this locality (see Fig. 7).

Table 2. Distribution of metacentric chromosome phenotypes between morphs of *Ilyodon* from the Río Terrero.

Morph	Number of metacentrics[2] 0	1	2	3	4	N per morph
A[1]	10	4	2	1	0	17
B[1]	0	5	16	10	6	37
					Total	54

$G_H = 35.6$ d.f $= 4$
$p << 0.005$*

[1] A = narrow mouth morph, B = broad mouth
[2] Note that except for the 'O' class, the number of metacentrics is a phenotype, not a genotype. Since homologs cannot be identified, the other metacentric classes may be genetically heterogeneous

The chromosomal polymorphism within the dichotomous *Ilyodon* from the Río del Tule is apparently part of a totally unprecedented 'step cline' in the number of metacentrics within the nondichotomous *Ilyodon* populations of the Río Coahuayana (Fig. 9, 10). Downstream populations have 2 metacentrics, while upstream ones have 10–12. A variety of evidence is consistent with the hypothesis that the chromosomal rearrangements involved have been pericentric inversions. The detailed distribution of chromosome phenotypes within the Río Coahuayana, including the existence of an intermediate population, will be presented elsewhere.

An overview of the relationship between the morphs

Our interpretation of the relationships of the *Ilyodon* morphs in the Río Coahuayana basin is as follows:

1) There is regular gene flow between the morphs in the Río del Tule and they are therefore, by definition, conspecific. The genetic basis of the morphological differences between the morphs are presently unknown. Some A × B matings probably result in progeny that are intermediate in phenotype.

2) The differences between the dichotomous (intermediates essentially absent) and the nondichotomous populations are probably caused by disruptive selection eliminating intermediates in the dichotomous populations. The nature of the morphological differences between the morphs suggests that food resources are primarily involved. An ecophenotypic component to the variation cannot now be excluded.

3) In at least one site in the Río del Tule, disruptive selection produces chromosomal differentiation between the morphs. Disruptive selection of this intensity, with proximate but differentiated interbreeding populations, is well known in other biological systems (review in Tavormina 1982). The hypothesis that disruptive selection is the primary factor responsible for the dichotomous populations does not necessarily rule out the existence of either some degree of assortative mating within morphs or the evolution of specialized genetic systems (e.g. supergenes), or an ecophenotypic component to the morphological variation.

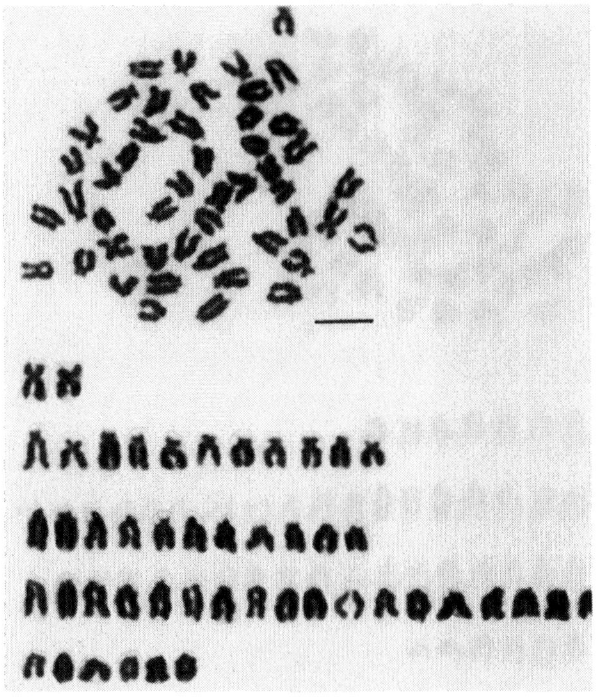

Fig. 9. Chromosomes and karyotype of a male *Ilyodon* collected at El Cobono (locality 11). Note the 2 metacentric chromosomes. This karyotype is characteristic of all *Ilyodon* sampled in the Río Salado and Barreras systems of the Río Coahuayana basin, but not of the Río Tuxpan (see Fig. 10).

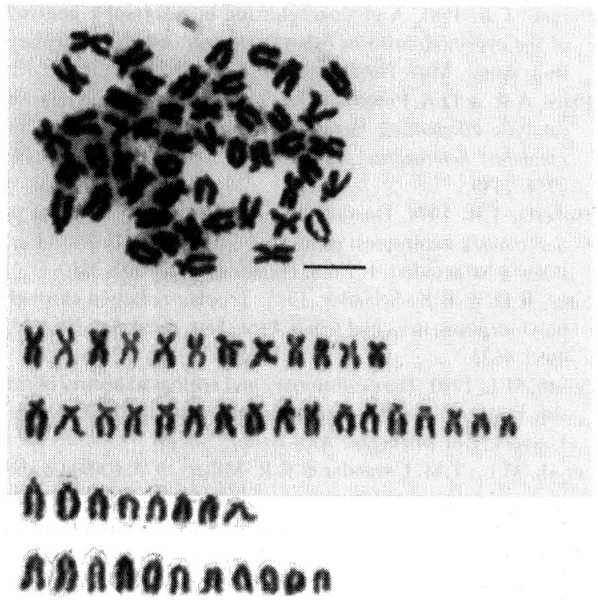

Fig. 10. Chromosomes and karyotype of a male *Ilyodon* from a laboratory line from near Cd. Tamazula (locality 20). Note 12 metacentric chromosomes. This karyotype is typical of nearly all localities surveyed in the Río Tuxpan system, except Antenquique (locality 18, with 10 metacentrics), Río Tonila (locality 15, with 6 metacentrics) and the Río del Tule – Río Terrero.

Conclusions

The relationships of the *Ilyodon* morphs in the Río del Tule can be compared to the species flocks of the family Cichlidae in the Great Lakes of Africa.

The *Ilyodon* morphs have all the attributes of 'good' species. They are recognizable by an array of correlated morphological characters, similar to those which distinguish closely related cichlid species. Moreover, though the *Ilyodon* morphs share a chromosomal polymorphism, they have differences in the modal number of metacentrics. If only a few individuals were karyotyped (a common practice) they might have appeared to be chromosomally divergent as well. But, the bulk of available evidence strongly suggests that they are conspecific. Thus, in the cichlid assemblages, the number of biological species, as delineated by trophic and color differences, may have been significantly overestimated. This consideration is not merely one of correct alpha taxonomy. The Rift Valley Lakes of Africa may contain sympatric, morphologically and ecologically differentiated forms that nonetheless are in regular genetic contact. This contact may be a significant component of their biology. The *Ilyodon* data suggest that the general level of divergence of conspecific fishes within even a single stream, much less an entire river system, cannot be deduced from morphological data alone.

In attempting to resolve the relationship of the *Ilyodon* morphs, and of the dichotomous to the nondichotomous populations, we have uncovered several cases of extreme genetic variation, within a single river system, in an abundant fish species with a seemingly continuous distribution. The biological factors responsible for this variation are as yet unknown.

Acknowledgements

We thank numerous colleagues for critical discussions, especially M. Bell, B.L. Brett, B. Chernoff, J. Humphries, I. Kornfield and M.L. Smith. Most especially, we extend cordial thanks to R.R. Miller for both his original explorations of the Río Coahuayana fish fauna and for his initial skepticism of our ideas on the relationships of the *Ilyodon* morphs; we have profited from both. The manuscript was improved by comments from R. Andrews and B. Wallace. The continuing support of the National Science Foundation, through grants DEB 76-20958 and DEB 79-23009, is gratefully acknowledged. Field work in Mexico was made possible by collecting permits issued by the Instituto Nacional de Pesca, Direccion General de Regulacion Pesquera, courtesy of Dra. Edith Polanco-Jaime.

References cited

Alverez, J. 1970. Mexican fishes. Secretaria de industria y comercio, Comision nacional consultiva de pesca, Mexico City. 166 pp. (In Spanish).

Barbour, C.D. 1973. A biogeographical history of *Chirostoma* (Pisces: Atherinidae): a species flock from the Mexican plateau. Copeia 1973: 533–556.

Barbour, C.D. & R.R. Miller. 1978. A revision of the Mexican cyprinid fish genus *Algansea*. Misc. Publ. Mus. Zool. Univ. Michigan 155: 1–72.

Eanes, W.F. & R.K. Koehn. 1978. An analysis of genetic structure in the monarch butterfly, *Danaus plexippus* L. Evolution 32: 784–797.

Fitzsimons, J.M. 1972. A revision of two genera of goodeid fishes (Cyprinodontiformes, Osteichthyes) from the Mexican plateau. Copeia 1972: 728–756.

Fryer, G. & T.D. Iles. 1972. The cichlid fishes of the great lakes of Africa. Oliver & Boyd, Edinburgh. 641 pp.

Greenwood, P.H. 1965. Environmental effects on the pharyngeal mill of the cichlid fish *Astatoreochromis alluaudi*. Proc. Linn. Soc. Lond. 176: 1–10.

Greenwood, P.H. 1981. Species flocks and explosive evolution. pp. 61–74. *In*: P.H. Greenwood & P.L. Forey (ed.) Chance, Change and Challenge – The Evolving Biosphere, Cambridge University Press & British Museum (Nat. Hist.), London.

Grier, H.J., J.M. Fitzsimons & J.R. Linton. 1978. Structure and ultrastructure of the testis and sperm formation in goodeid teleosts. J. Morph. 156: 419–438.

Hubbs, C.L. & C.L. Turner. 1939. Studies of the fishes of the order Cyprinodontes. XVI. A revision of the Goodeidae. Misc. Publ. Mus. Zool. Univ. Michigan 42: 1–80.

Johnson, M.S. 1971. Adaptive lactate dehydrogenase variation in the crested blenny, *Anoplarchus*. Heredity 27: 205–226.

Kingston, D.I.L. 1979. Behavioral and morphological studies of the goodeid genus *Ilyodon*, and comparative behavior of the fishes of the family Goodeidae. Ph.D. Thesis, University of Michigan, Ann Arbor. 462 pp.

Kornfield, I., D.C. Smith, P.S. Gagnon & J.N. Taylor. 1982. Trophically divergent cichlid fish are one biological species. Evolution 36: 658–664.

Maynard Smith, J. 1966. Sympatric speciation. Amer. Nat. 100: 637–650.

Miller, R.R. & J.M. Fitzsimons. 1971. *Ameca splendens*, a new genus and species of goodeid fish from western Mexico, with remarks on the classification of the Goodeidae. Copeia 1971: 1–13.

Miller, R.R. & T. Uyeno. 1980. *Allodontichthys hubbsi*, a new species of goodeid fish from southwestern Mexico. Occ. Pap. Mus. Zool. Univ. Michigan 642: 1–13.

Mitton, J.B. & R.K. Koehn. 1975. Genetic organization and adaptive response in allozymes to ecological variables in *Fundulus heteroclitus*. Genetics 79: 97–111.

Myers, G.S. 1960. The endemic fish fauna of Lake Lanao, and the evolution of higher taxonomic categories. Evolution 14: 323–333.

Parenti, L.R. 1981. A phylogenetic and biogeographic analysis of the cyprinodontiform fishes (Teleostei, Atherinomorpha). Bull. Amer. Mus. Nat. Hist. 168: 335–557.

Place, A.R. & D.A. Powers. 1979. Genetic variation and relative catalytic efficiencies: lactate dehydrogenase B allozymes of *Fundulus heteroclitus*. Proc. Nat. Acad. Sci. USA 76: 2354–2358.

Roberts, T.R. 1974. Dental polymorphism and systematics in *Saccodon*, a neotropical genus of freshwater fishes (Parodontidae, Characoidei). J. Zool. (London) 1973: 303–321.

Sage, R.D. & R.K. Selander. 1975. Trophic radiation through polymorphism in cichlid fishes. Proc. Nat. Acad. Sci. USA 72: 4669–4673.

Smith, M.L. 1980. The evolutionary and ecological history of the fish fauna of the Río Lerma basin, Mexico. Ph.D. Thesis, University of Michigan, Ann Arbor. 205 pp.

Smith, M.L., T.M. Cavender & R.R. Miller. 1975. Climatic and biogeographic significance of a fish fauna from the late Pliocene-early Pleistocene of the Lake Chapala basin (Jalisco, Mexico). Univ. Michigan Pap. Paleontol. 12: 29–38.

Sokal, R.R. & F.J. Rohlf. 1981. Biometry. 2nd ed. W.H. Freeman & Co., San Francisco. 859 pp.

Spiess, E.B. 1977. Genes in populations. John Wiley & Sons, New York. 779 pp.

Tauber, C.A. & M.J. Tauber. 1977. A genetic model for sympatric speciation through habitat diversification and seasonal isolation. Nature 268: 702–705.

Tavormina, S.J. 1982. Sympatric genetic divergence in the leaf-mining insect *Liriomyza brassicae* (Diptera: Agromyzidae). Evolution 36: 523–534.

Templeton, A.R. 1980. Modes of speciation and inferences based on genetic distances. Evolution 34: 719–729.

Turner, B.J. & D.J. Grosse. 1980. Trophic differentiation in *Ilyodon*, a genus of stream-dwelling goodeid fishes: speciation versus ecological polymorphism. Evolution 34: 259–270.

Vrijenhoek, R.C. 1978. Coexistence of clones in a heterogeneous environment. Science 199: 549–552.

Wourms, J.P. 1981. Viviparity: the maternal-fetal relationship in fishes. Amer. Zool. 21: 473–515.

Wright, S. 1978. Evolution and the genetics of populations. Vol. 4. Variability within and among natural populations. Univ. of Chicago Press, Chicago. 580 pp.

Originally published in Env. Biol. Fish. 9: 159–172

Natural and sexual selection on color patterns in poeciliid fishes

John A. Endler
Department of Biology, University of Utah, Salt Lake City, UT 84112, U.S.A.

Keywords: Crypsis, Mate choice, Mimicry, Neotropical fishes, Predation, Predator avoidance, Sexual dichromism, Sexual dimorphism

Synopsis

In poeciliid fishes, sexual dichromism is associated with larger size and larger broods, but there is no relationship between sexual size dimorphism and sexual dichromism, or between degree of dichromism and color pattern polymorphism. Factors are discussed which influence the evolution of color pattern polymorphisms, sexual dimorphism and dichromism. Detailed studies of South American species have shown that the color patterns of poeciliid fishes have predictable effects in (1) avoiding diurnal visually hunting predators; (2) mating success; and (3) species recognition. Data from some Central American species indicate that some color pattern elements may be closely linked to physiologically variable loci, which further affect the variation in color patterns. Different elements of any given color pattern can be influenced by different modes of natural selection; in guppies the relationship between predation intensity and color pattern is different for melanin, carotenoid, and structural colors. Different color patterns have different degrees of conspicuousness on different backgrounds, and may appear differently to predators and mates with differing visual abilities.

Introduction

Natural selection is a major component of the theory of evolution, yet little is known about its function or magnitude (Lewontin 1974, Endler 1980). Sexual selection is a subject of great interest because it may result in genetic change which has little to do with environmental factors (Lande 1980, 1981, Kirkpatrick 1982). The best examples of both phenomena are in studies of color patterns because they are relatively easy to measure, and their adaptive significance is usually obvious (Endler 1978, 1980). Discussion of natural and sexual selection will be limited to the Poeciliidae; data from other families are very scanty (see Endler 1978).

The teleost family Poeciliidae is distributed from Delaware and southern Arizona through northeastern Argentina and Ecuador (Rosen & Baily 1963). It is found in most fresh and brackish water habitats, but tends to be absent from very fast flowing mountain streams, large rivers, and far inland. Poeciliids are primarily fish of shallow waters, and except for the piscivorous *Belonesox*, are omnivorous.

Sexual dimorphism and dichromism

Male poeciliids are usually smaller than females; the ratio of male to female maximum observed sizes (TL) averages 0.7, and ranges from 0.4 in *Poecilia scalpridens* and *Poeciliopsis gracilis* to 0.9 in *Neoheterandria cana* and *Priapella bonita*, and about 1.0 in *Poecilia petenensis* (data from Jacobs

Thomas M. Zaret (ed.), Evolutionary ecology of neotropical freshwater fishes. ISBN 90 6193 823 6

1971). Sexual dimorphism in color patterns (dichromism) is also highly variable in the family, although some genera tend to be more (*Poecilia*, *Xiphophorus*) or less (*Gambusia*, *Girardinus*, *Poeciliopsis*) dichromic than the family as a whole. The female color patterns are generally simpler and less developed than in males. If we split the species into those which are strongly, weakly, or not sexually dichromatic, there appears to be no relationship between the degree of size dimorphism and the degree of dichromism [$F(2,123) = 0.4$, $P \gg 0.05$, data from Jacobs 1971].

Ghiselin (1974) suggests that in species with small males, females invest energy in producing eggs and males invest energy in seeking out females rather than in gonadal development. The differences in size may arise because energy for growth is shunted into swimming and sexual displays in males, while female energy allocation is towards growth of body and gonads, egg production (in terms of total weight of eggs), and food transfer to developing embryos (Balon 1975, 1983, Wourms 1981), which is proportional to female size. There is a significant correlation between female size (TL) and brood size among species [$r = 0.35$, $P < 0.02$, data of Jacobs 1971], and within a species (Reznick & Endler 1982). Male poeciliids spend much more time on reproductive activity than females (Baerends et al. 1955, Liley 1966, Constantz 1975 & unpublished manuscript), probably because there are very few females receptive and fertile at any one time and place (Liley 1966, Seghers 1973, Balsano et al. 1981). Gestation time may be an additional indicator of female 'investment', but curiously enough, it is positively correlated with the male/female size ratio [$r = 0.40$, $P < 0.02$], the ratio is closer to 1.0 in species which take longer to mature their broods.

Baylis (personal communication 1982) has suggested an alternate hypothesis to explain why males are usually smaller than females, based upon the energetics of livebearing. Poeciliids have greatly reduced fecundity relative to other fishes by bearing offspring and supplementing embryonic yolk with maternal nutrients (Balon 1975, 1983, Wourms 1981). The advantage of a longer gestation time is a higher 'quality' offspring – able to eat larger food items, better able to maneuver in streams, avoiding predators and rapidly flowing water. Larger females produce larger broods. But there are energetic and predator risk costs to bearing young for long periods. Smaller females may be more efficient in bearing young (lower somatic maintenance costs), so it is possible that smaller females would be favored in areas with either low food availability or very high predation. This predicts less size dimorphism in species and areas with high predation compared to low predation, and in areas with little or no predation, the fecundity-size relationship favors larger size (Baylis, personal communication 1982). This appears to be true for two species: In *Poecilia reticulata* the ratio of male size at maturity to minimum gravid female size is 1.01 in areas of high predation and 0.94 in areas of weak predation (Reznick & Endler 1982). In *Phalloceros caudimaculatus* the mean adult size ratios are 0.98 ± 0.12 and 0.98 ± 0.08 in dangerous and intermediate predation areas, 0.87 ± 0.12 and 0.82 ± 0.05 in arthropod-only and safe areas (Endler 1982). Baylis's predictions are independent of sexual selection, so it is consistent that there was no correlation between sexual dimorphism and dichromism among species. An additional reason for smaller females under high predation is that larger females take longer to mature. If the mean date of maturity approaches or exceeds the mean life expectancy, then this favors earlier maturity. This predicts smaller size at maturity in areas with high predation, which has been found in guppies (Reznick & Endler 1982) and *P. caudimaculatus* (Endler 1982). Males are also smaller at maturity in areas with high predation, but their size does not decline as rapidly with increased predation, hence the smaller size dimorphism in areas with high predation. There are presently insufficient data to know whether this applies among species.

There is a significant relationship between the degree of sexual dichromism and the number of young per brood [$F(2,49) = 3.75$, $P < 0.05$, data of Jacobs 1971]; species which have a greater difference between male and female coloration tend to have a larger brood size ($r = 0.39$). In addition, the strongly dichromic species also tend to be those with larger female size; sexually monochromic or weakly dichromic species tend to be small [$F(2,123) = 4.62$, $P < 0.025$]. It is possible that these relationships result from differences in predation intensity among species. The stronger the dichromism the brighter the males, which means greater risk to predation (Fisher 1930, Cott 1940, Haskins et al. 1961). In addition, larger species are preyed upon by larger more dangerous predators than smaller species. Those individuals with larger broods (larger individuals), have a greater chance of leaving reproducing offspring than those with genetically smaller broods, and therefore will be more successful in places with dangerous predators. This may explain the size variation among species and does explain the variation among populations of *P. caudimaculatus* and guppies living with differing predator communities. In guppies the differences are heritable (Reznick 1980, Reznick & Endler 1982). Dichromism may result in relatively more predation on males than on females. Females which are able to suppress the expression of color pattern genes will be at an advantage to those which do not, favoring dichromism. Males cannot suppress their colors or their mating will be impaired, either through poorer species recognition or sexual selection. It is interesting that many of the known color pattern genes are sex limited (color pattern genes not expressed in normal females) and sex-linked in sexually dichromic species which have been examined. The correlation between dichromism and brood size may have arisen because both may be affected independently but in parallel by predation.

Color pattern polymorphisms in poeciliids

Color patterns may be polymorphic (variable among individuals) in sexually dichromic or monomorphic species, for example *P. reticulata* and *Phalloceros caudimaculatus*, respectively. Color pattern polymorphisms are uncommon in the family as a whole, but are relatively common in the genera *Poecilia* and *Xiphophorus*. *P. reticulata* is one of the most color polymorphic vertebrates, and one of the first to be studied genetically. The genetics of the color pattern polymorphisms of *P. reticulata* and *Xiphophorus* species are fairly well known and are similar, though much more complex in *P. reticulata* (Kallman 1975, Yamamoto 1975). Genes which control the presence or absence of particular colored spots are usually sex-linked and often sex-limited. As with sexual dichromism, color pattern polymorphism tends to be found in larger species. Polymorphism may in fact be more common and well developed in the family than appears from the literature, merely because most species have not been examined, for example, *Phalloceros caudimaculatus* (Endler 1982).

Color patterns serve three major functions in animals: (1) thermoregulation, (2) reduction of predation, and (3) intra- and inter-specific communication (Cott 1940, Endler 1978). Although dark coloration is known to increase temperature in some cold water copepods and mosquito larvae (Byron 1981, L.T. Nielson, personal communication 1982), this has not been studied in fish, and warm water may minimize the importance of colors in the heat balance of neotropical fishes. It is interesting that the black forms of *Gambusia* are found in the north of the distribution of the genus, though they are not necessarily correlated with temperature regimes within species (Martin 1977). Color patterns may serve to hide fish from predators (crypsis, Cott 1940, Wickler 1968, Endler 1978), or to make distasteful or mimetic fish more conspicuous (aposematism, Cott 1940, Wickler 1968, Keenleyside 1979). Color patterns or pattern elements are used by many species during mate recognition, courtship, and in other social contexts (Baerends et al. 1955, Liley 1966, Rosen & Tucker 1961, Constantz 1975). Variation in color patterns will affect their efficiency in one or more of these functions, so color pattern polymorphisms allow us to study the function and adaptation of color patterns.

A complicating factor in the study of color pattern polymorphism, and in fact all polymorphisms, is the possible effects of linked genes with strong affects on fitness, and of pleiotropic effects of particular loci affecting fitness in unexpected ways. Examples are found in *Xiphophorus* species. Kallman & Borkoski (1978) found a sex-linked gene *P* which controls growth and the onset of maturity in *X. maculatus* because it was initially linked to a color pattern gene. Later investigations revealed the effects of other loci (Schreibman & Kallman 1978). *X. helleri* (Peters 1964), *X. maculatus* (Borowsky 1981), *X. montezumae* (Zander 1965), *X. pygmaeus* (Rosen & Kallman 1969), and *X. variatus* (Borowsky 1978) show a dimorphism in male size and color pattern development which may be associated with this gene complex. It is unclear in many cases whether the differences in size and color patterns among the different male types result from close linkage with *P* genes, or pleiotropic effects of the color pattern genes themselves. Except for Kallman, no one has looked for recombinants. These effects complicate studies of the functions of color patterns because it is difficult to disentangle the direct and indirect effects of each allele.

Some color pattern variation may be a function of, or even result in, differences in social status. For example, compared to subordinates, dominants of the following species are characterized by: young *P. reticulata* - black iris (Martin & Hengstabuck 1981); male *Poeciliopsis occidentalis* - darker (Constantz 1975); male *Gambusia heterochir* - paler (Warburton et al. 1957); male *X. variatus* - yellowish-red (Borowsky 1973). Such differences may or may not be influenced by a genetic polymorphism, and this must be considered in all polymorphism studies.

Color pattern polymorphisms may be found only in males (for example, *P. reticulata*), or in both sexes (*Ph. caudimaculatus*); sexual dichromism is independent of color pattern polymorphism within the family. This makes interpretation of the effects of predation on dichromism and polymorphism complicated. For example, as predation increases, we expect color patterns to be less conspicuous as a result of a decline in the frequency of conspicuous color pattern elements. But it is not clear if there should also be a decline in dichromism with increased predation. The relationship between sexual dichromism, crypsis, and predation may not be obvious, and may not necessarily run the same way. At intermediate predation intensities, sexual dichromism may be stronger than at very high or very low predation: At low predation there is little selection for crypsis in either sex. At high predation, selection for inconspicuousness may be so strong that suppression of certain color pattern elements in females together with selection against conspicuous pattern elements in males results in very small differences in color patterns between sexes. At intermediate predation intensity, suppression of colors in females encourages dichromism, but the balance between sexual selection and predation encourages polymorphisms. Of course the direct effects of sexual selection may entirely obscure this relationship. The interrelationships between dichromism and color pattern polymorphism need much more work.

Color pattern polymorphism in *Poecilia reticulata*

Guppies, *Poecilia reticulata* Peters, show a complex color pattern polymorphism, and enough is known about their ecology, behavior and genetics to make significant progress in understanding the polymophism (Haskins et al. 1961, Farr 1976, 1977, 1980a, b, Yamamoto 1976, Endler 1978). Natural populations in Trinidad and northeastern Venezuela are so polymorphic that no two males are alike (Fig. 1a-c). The color patterns consist of a mosaic of patches varying in color, size, position and reflectivity, and are controlled by many X- and Y-linked genes. These genes are expressed only in adult males, and female genotypes can be revealed by testosterone treatment (Haskins et al. 1961).

The color patterns in a particular population represent a balance between selection for crypsis by predators and selection for conspicuousness by sexual selection (Fisher 1930, Haskins et al. 1961, Gandolfi 1971, Greene 1972, Farr & Herrnkind 1974, Farr 1976, 1977, Gorlick 1976, Endler 1978, 1980). Two exceptions are the studies by Farr (1980a) on sexual selection, and by Seghers (1973) on predation, in which color patterns had no effect. A significant problem in these and most other studies is the lack of control for the visual background against which mates or prey are seen, and also the difficulty in defining what was and was not conspicuous to predators and mates. A color pattern on a bare laboratory background can be quite conspicuous, yet be quite cryptic against the gravel backgrounds of their native streams (Endler 1978). One guppy which seems more conspicuous than another in the laboratory may actually be less conspicuous in the field. In the studies described below, background color patterns were taken into account.

In order to be cryptic (inconspicuous), a color pattern must represent a random sample of the background normally seen by visually hunting predators at the time and place at which the prey are most vulnerable to predation (Endler 1978). Any deviation of the whole pattern from the background in the distributions of patch size, color, or brightness, will make the color pattern conspicuous, and the degree of conspicuousness is proportional to the deviation between animal and background distributions. In areas of intense predation, the background match should be better than in areas of weak predation. On the other hand, sexual selection favors color patterns which deviate from the background. Sexual selection and predation occur together in natural guppy populations, so a compromise is achieved which depends upon the relative intensities of the two processes. The compromise will be different for pattern elements which certain predators cannot see, and if the dominant predator has very different vision than we do, our estimate of conspicuousness may be wrong or misleading. Similarly, if predator's and prey's vision differ, the same pattern may differ in conspicuousness for predators and mates (Endler 1978).

Predation intensity in natural guppy populations

Within the natural geographical range of guppies, it is possible to choose localities where the diurnal visually hunting predators are common and well defined (Haskins et al. 1961, Endler 1978, 1980). By careful choice of localities, one can eliminate the effects of diurnal fish-eating insects, snakes and birds (see Endler 1978 for details). The aquatic diurnal visually hunting predators (fish and crustaceans) are easily censused because the study streams are clear, have gravel bottoms with no aquatic vegetation, and owing to the lack of human activity, the animals are relatively tame. The predators are: *Rivulus hartii* (Cyprinodontidae), *Aequidens pulcher*, *Cichlasoma bimaculatum* and *Crenicichla alta* (Cichlidae), *Hemibrycon dentatum* and *Astyanax bimaculatus* (Characidae), and the freshwater prawn *Macrobrachium crenulatum* (Palaemonidae). These are found in the south-draining streams of the Venezuelan Paria Penninsula and Margarita Island, and the Northern Range of Trinidad (Endler 1978). A second predator fauna is found in the north-draining streams in these areas, also at the narrow eastern extremity of the Paria Penninsula, and Tobago Island: *R. hartii* (as before), *Agonostomus monticola* and possibly *A. microps* (Mugilidae), *Eleotris pisonis* and an unidentified *Eleotris* (Gobiidae), occasionally *Centropomis unidecimalis* (Centropomidae), occasionally *Gobiomorus dormitor* or *Dormitator maculatus* (Gobiidae), *Macrobrachium crenulatum* and *M. faustinum* (Palaemonidae). Because the first fauna consists primarily of mainland families and the second mostly of marine families, I will refer to the two predator faunas as the Mainland and Caribbean faunas, respectively.

The predators are distributed in a number of characteristic assemblages as a result of habitat choice and requirements, predation by other species, and chance events. In the Mainland fauna the assemblages and their codes are: (R) *R. hartii* alone; (M) *R. hartii* + *M. crenulatum*; (A) *A. pulcher* alone or with *R. hartii* uncommon or rare; (K) Characins: *A. bimaculatus and/or, H. dentatum*; (A + K) *A. pulcher* and characins together; (C + A + K) *C. alta*, *A. pulcher* and characins together, occasionally *Cichlasoma bimaculatum* (Endler 1978). In the Caribbean fauna: (R + M) as M, but often with *M. faustinum* in place of *M. crenulatum*; (Ag) *A. monticola* and occasionally *G. dormitor*; (E) *Eleotris* species with *Agonostomus* species and occasionally *C. unidecimalis* or *D. maculatus*. Both sets of assemblages are roughly in order of decreasing elevation. As in many other freshwater stream systems (Hynes 1970, Whitten 1975), the number of species and the number of predators increases downstream in both faunas.

From stomach-content analysis of several populations and 39 h of direct observation of two populations in the upper El Cedro river, Trinidad (Endler 1978, and unpublished) it is possible to rank the Mainland predators by increasing danger to guppies: *R. hartii* < *A. pulcher* = *C. bimaculatus* < *H.*

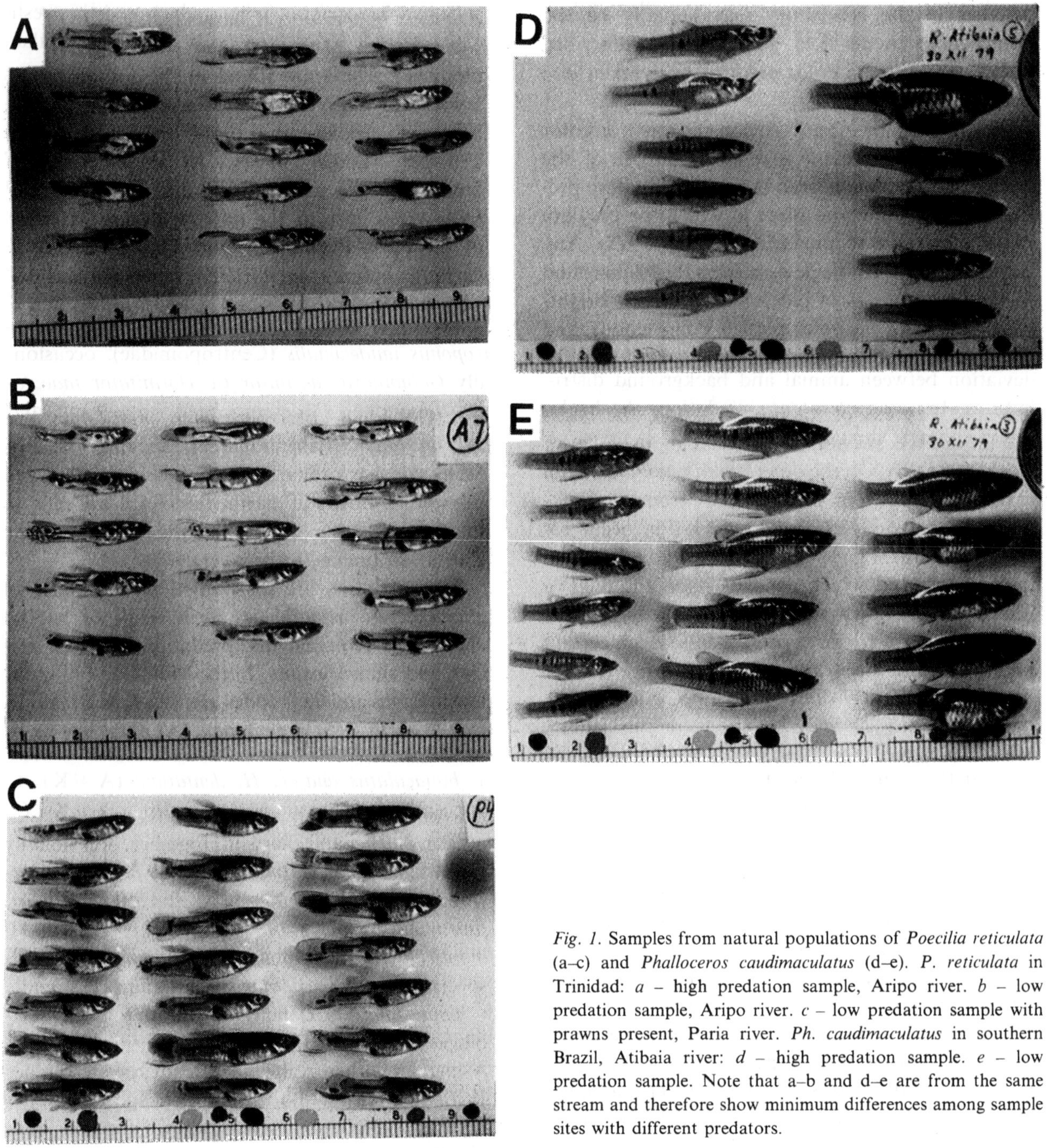

Fig. 1. Samples from natural populations of *Poecilia reticulata* (a–c) and *Phalloceros caudimaculatus* (d–e). *P. reticulata* in Trinidad: *a* – high predation sample, Aripo river. *b* – low predation sample, Aripo river. *c* – low predation sample with prawns present, Paria river. *Ph. caudimaculatus* in southern Brazil, Atibaia river: *d* – high predation sample. *e* – low predation sample. Note that a–b and d–e are from the same stream and therefore show minimum differences among sample sites with different predators.

dentatum < *A. bimaculatus* ≪ *C. alta*. The attack rate on guppies per hour was about 3 for *C. alta*, 1 for the characins, and less than $\frac{1}{2}$ for *A. pulcher* and *R. hartii*. Casual observations of *M. crenulatum* suggest that it attacks at least as often as *R. hartii*. Aquarium observations indicate that prawns are more dangerous predators than *R. hartii*; they will eat all guppies in a 38 l tank in much less time than does a *R. hartii*. On the basis of stomach contents, the Caribbean predators may be ranked by increasing danger: *R. hartii* < *A. monticola* = *G. dormitor* << *E. pisonis*. *A. monticola* is similar to *A. bimaculatus* and *E. pisonis* is similar to *C. alta* in danger to guppies.

Because the predator assemblages vary in the number of predators and their degree of danger to guppies, it is possible to choose various points on a visual selection intensity gradient; as one moves downstream and among valleys, one shifts from few species which rarely eat guppies to many species, some of which commonly eat guppies. Thus we can rank the predator assemblages by increasing visual selection intensity: Mainland A ≤ R ≤ M < K < A + K < C + A + K, and Caribbean R + M < Ag < E. The streams also vary in background, and there is a tendency for the more dangerous predator assemblages to be found in places with sand and small gravel, but it is possible to control for background among predator assemblages by careful choice of sampling sites.

Guppy color pattern polymorphisms and predation intensity

The a priori knowledge of visual predation intensity gradients provides an opportunity to test the theory of color pattern and background matching. As visual selection intensity increases, we expect the balance between crypsis and sexual selection to shift. In places with little predation the effects of sexual selection should predominate, favoring relatively conspicuous coloration. In places with intense predation, the effects of predation should be stronger than sexual selection, favoring relatively cryptic coloration. This is exactly what appears to be happening. As predation intensity increases, the number and size of patches decreases (Fig. 2, 3). The differences and trends for most colors are highly significant (Endler 1978). The new data from the Caribbean predator fauna (Fig. 3) are very similar to the Mainland results (Fig. 2).

The reduction in number of patches with increasing predation is primarily a result of the decrease in frequency of structural colors (blue, iridescent, bronze). Structural colors work on refraction, interference and differential reflection, and are therefore much more conspicuous than the pigment colors, which work on differential absorption. In addition, structural colors tend to reflect maximally at a particular angle, and flash during movement of the fish. This can be very conspicuous from a long distance and hence especially bad in a high predation area. There are several reasons why the size of spots should decrease with increasing predation: (1) the background color pattern patch size tends to be smaller at lower elevations, where more dangerous predators are found; (2) a smaller patch will be below the visual acuity angle threshold or the color detection angle threshold more often than a larger spot; (3) the distance to first detection will be larger for a smaller than a larger spot. For further details, see Endler (1978).

Experiments on predation, and the effects of backgrounds

The concordance between the Mainland and Caribbean fauna results is strong inferencial evidence that predation has a significant effect on guppy color patterns. Experimental manipulation is a direct method of demonstrating that the color patterns at a particular place are an evolutionary response to sexual selection and predation. Experiments were performed in artificial streams in a greenhouse and by field transfer experiments.

Greenhouse experiments. — A large outbred population was derived from 18 localities in Trinidad and Venezuela and distributed among ten segments of an artificial stream system in a temperature controlled (25° C) greenhouse. Four of the segments were made as physical full scale replicas of a typical *Crenicichlia alta* territory (C pools), and the remaining six were replicas of streams containing

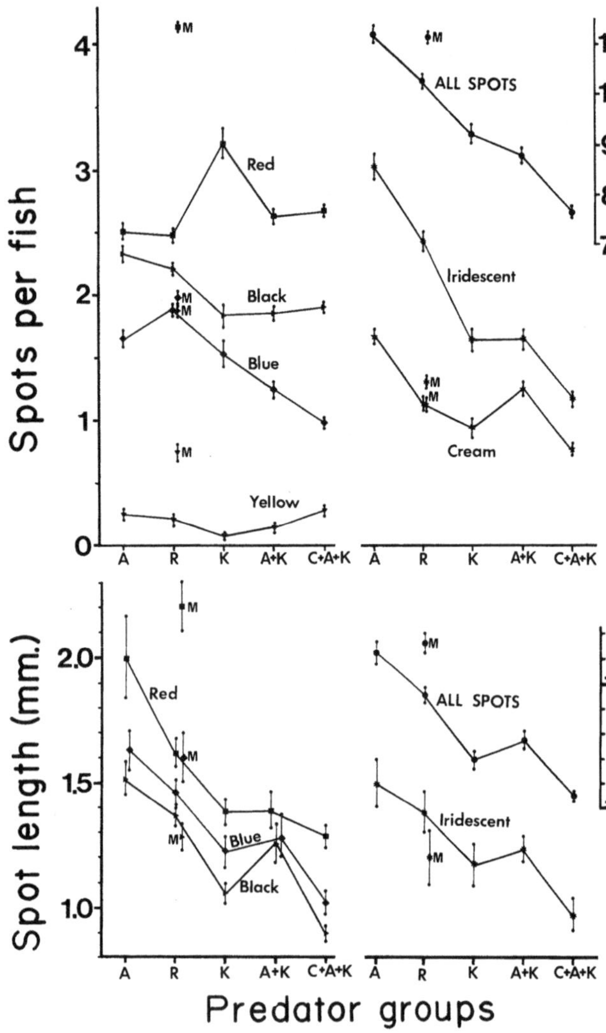

Fig. 2. Numbers of spots per fish and spot length as a function of rank predation intensity in the Mainland predator fauna. The predator groups are ranked in order of increasing predation intensity. For details see text. M stands for localities with freshwater prawns and *Rivulus hartii* rare. The vertical bars include two standard errors around the mean. Note the different scale for all spots pooled (from Endler 1978).

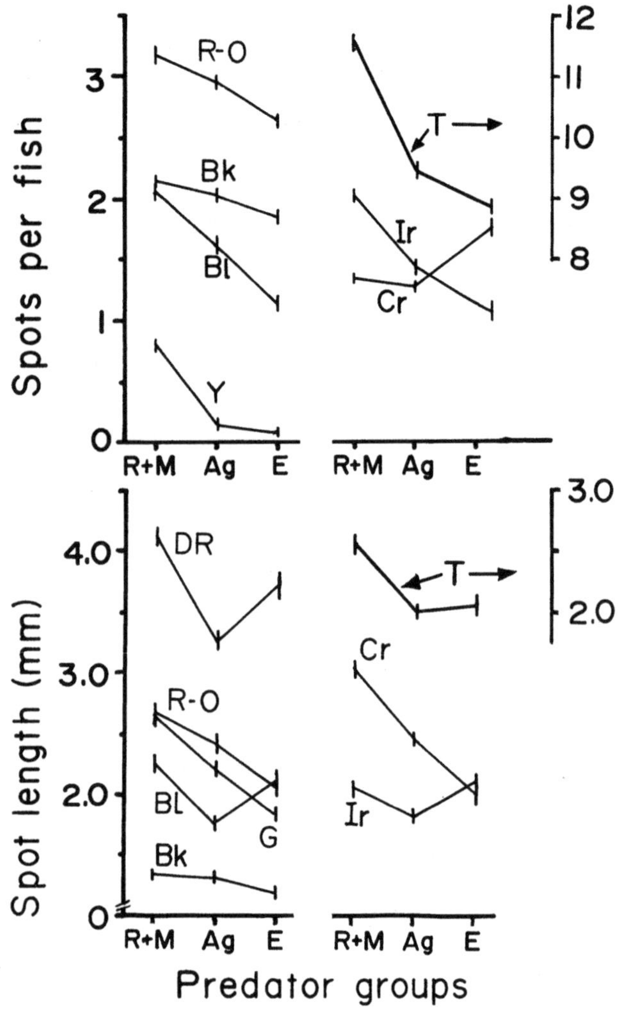

Fig. 3. Numbers of spots per fish and spot length for the Caribbean fauna. The predator groups are ranked in order of increasing predation intensity, as in Figure 2. Color symbols: R-O – red-orange, DR – dark red, Bk – black, Bl – blue, G – green-bronze, Cr – cream-white, Ir – silver- iridescent, Y – yellow, T – all spots. All show significant trends except for Bk, Cr spot number, and DR, Bl, Ir spot size. Note the separate scale for all spots pooled.

Rivulus hartii (R pools). Half of the C and half of the R pools had coarse gravel, and the remaining pools had fine gravel, with the same color and brightness frequency distributions. After six weeks, a *C. alta* adult was placed in each of the four C pools, six *R. hartii* were introduced into each of four of the R pools, and the remaining two R pools had no predators. This is a 3 × 2 factorial design, testing simultaneously for the effects of background patch size (fine and coarse) and predation intensity (none, low and high). Complete censuses of all fish were taken at 5 and 14 months (about 3 and 9 generations) after the predators were introduced. For further details, see Endler (1980).

The null hypothesis is that there will be no effect either of predation intensity or background color patch size.

The color pattern characteristics of each predation intensity treatment diverged significantly from the starting values, converged remarkably on the field results, and there was a significant effect of background (Fig. 4). There were smaller and fewer spots per fish in high compared to low predation, and this was primarily due to a reduction in the number of structural colored spots and a reduction in the size of the pigment color spots. In the two predation intensity levels, spot size was larger on large gravel and smaller on small gravel. The similarity to the actual background grain was greater in the pools with *C. alta* (high predation). Note that the effect of predation intensity was much greater than that of background (Endler 1980).

There were direct signs of sexual selection favoring males which contrast with the background rather than resemble it. First, in the control and *R. hartii* pools, there was a steady increase in the number and size of spots. Secondly, in the absence of predation the spot size became larger on small gravel than it became on large gravel (Fig. 4). Males can contrast with the background by having spots either larger than or smaller than the background grain, and the direction of the deviation seems to depend upon the initial difference between the guppy spot size and gravel size. Since these results were all highly significant (Endler 1980), it is clear that the color patterns in a particular place are affected by sexual selection and predation, with a complicated but predictable balance between the two and interaction with the background grain. Experiments are in progress which test for the effects of gravel color frequencies rather than patch (grain) size.

Field experiments. — Additional tests of the effects of predation and sexual selection were performed in the field. A smalll tributary of the Aripo River in Trinidad contained *R. hartii* but no guppies. Two hundred guppies were taken from the main stream, which contained *C. alta* and other predators (C + A + K) and introduced into the tributary (R). A second tributary nearby was physically and biologically very similar to the first, except that it had guppies; this served as a control (R). Thus guppies were moved from a place with intense predation to a place with little predation. A sample was taken from all three localities 23 months or about 14 generations later. The null hypothesis is that there should be no change in the introduced population, and the alternative hypothesis is that, under reduced predation, the effects of sexual selection should increase the conspicuousness of the introduced guppies. The null hypothesis was rejected; all color pattern characteristics diverged from the mainstream (ancestral) population and converged on those typical of low predation populations, including the control. The color patterns became more conspicuous by an increase in the size and numbers of spots and a shift towards structural

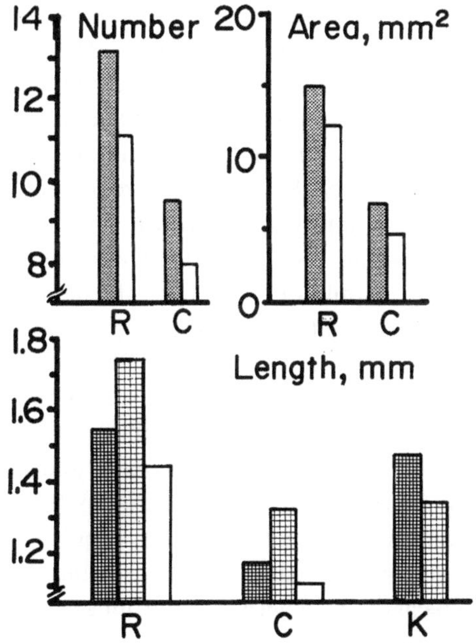

Fig. 4. Summary of the results of the greenhouse experiment: total number of spots, total area of spots and mean spot length per fish. C – high predation, R – low predation, K – no predation. Shaded bars are the greenhouse results and open bars the field results with the same rank predation intensity. Fine hatching – fine gravel. Coarse hatching – coarse gravel. Differences between treatments are highly significant. For further details see text.

colors (Endler 1980). The inverse experiments, introducing *C. alta* into low predation areas and following the color pattern changes in guppy populations, are in progress in two streams.

Background color pattern and female choice

What is conspicuous to a predator will also be conspicuous to a female. By the same arguments used for crypsis, we would expect that the success of males with particular color pattern characteristics will depend upon the background against which they are seen by females, and a given male will not be equally successful on all backgrounds.

The sexual behavior of guppies has been described in detail (Haskins & Haskins 1949, 1950, Clark & Aronson 1951, Baerends et al. 1955, Liley 1966, Farr & Herrnkind 1974, Farr, 1976, 1977, 1980a,b, Kennedy 1979). Males constantly search for receptive females. If a female remains relatively stationary, then the male will go through a series of displays in which his body is placed 30°–35° away from directly in front of her, and presented with fins spread and body oriented so that maximum surface is presented to one of her eyes. During part of the display (usually the S shaped 'sigmoid' part) his body is moved rapidly up and down and through a small angle along his long axis. The displays are followed by copulatory attempts if she remains stationary. Fertilization is internal, by means of a modified anal fin or gonopodium (Rosen & Gordon 1953). Copulation requires active cooperation by the female. Females copulate more often with males with conspicuous color patterns and higher courtship rates. During and immediately following copulation, males usually exhibit rapid 1–4 mm movements ('postcopulatory jerking'), and this appears to be associated with spermatophore transfer. Multiple insemination is common, but in general, the last male to mate prior to zygote formation is responsible for most of the zygotes (Winge 1937, Hildemann & Wagner 1954).

There is a second mode of male behavior called gonopodial thrusting (Baerends et al. 1955) or rape (Farr 1980b) in which a male tries to inseminate a female without first displaying to her. It is most common when females are not receptive, and appears to be independent of prior experience (Farr 1980b). If thrusting always resulted in insemination, then this would weaken the effects of female choice of males with particular color patterns. However, thrusting rarely results in successful genital contact, and the percentage of actual fertilization is even rarer (Clark & Aronson 1951, Baerends et al. 1955, Liley 1966). The evidence for successful fertilization is anecdotal, and in the published experiments, there was no opportunity for subsequent copulation with another male. In natural populations subsequent copulation is possible and probably frequent. Farr (1980b) mentions that thrusting can potentially lead to insemination during any part of the ovarian cycle, while standard courtship is presumably only successful at the appropriate period for fertilization in the cycle. But in the rare instance when a thrust actually results in a spermatophore transfer, the sperm may have to wait until the eggs are mature. If the female is inseminated during standard courtship after a thrust by another male, the new spermatophores will contain younger sperm (presumably more viable), a greater concentration of sperm, and they will be at an advantage compared to the older sperm because they were the last ones in. Thus, the probability that thrusting actually leads to a significant percentage of the fertilizations, hence of importance to the evolution of female choice, is very low. In addition, since the female has little opportunity to choose between different males attempting to thrust, its direct effect on the evolution of color patterns may be small. Much more work needs to be done on the actual fertilization success rate of gonopodial thrusting, and whether or not males which thrust can or cannot be chosen or resisted by females.

Sexual selection appears to be mediated through choice by females of males rather than choice by males or inter-male competition (references above, also Greene 1972, Gandolfi 1971, Gorlick 1976). Inter-male competition frequently occurs in small experimental aquaria (Gandolfi 1971, Farr 1976, 1977, 1980a,b, Gorlick 1976, Kennedy 1979), but is absent in very large aquaria at low densities, in the artificial greenhouse streams (Endler, unpublished), in small artificial streams (Liley, personal communication 1978), or in natural populations in

Trinidad (Endler unpublished, Liley personal communication 1978). The reasons for dominance hierarchies and inter-male aggression in spatially restricted artificial populations and in juveniles (Martin & Hengstebeck 1981), are unknown. During courtship display by two (or rarely more) males to a single female, each displays maximally to the female. If one male tried to interact with the other in some way, it would lose time in displaying to the female, and she might then chose the other male, or move off to another area. One does see males jockeying for best display position, but these are almost always cases in which the female is unreceptive, but being followed by several 'hopeful' males. It is arguable whether two males are trying to maximize their display to a receptive female, or take the female's attention away from another male. I will therefore regard the system as primarily one of female choice rather than inter-male competition.

In order to investigate the factors influencing female choice, guppies were taken from the R pools of the greenhouse population. Males were drawn at random, matched for size and age, and ranked with respect to number and size of (a) red and orange spots, (b) blue spots, and (c) all colors. The matched males were then presented to virgin females in the greenhouse with the following gravel backgrounds: (i) black & white; (ii) black, white, & red; (iii) black, white, & blue; (iv) multicolored bright. The fish were watched for an hour, recording their behavior. The preliminary results are quite interesting (Fig. 5) and show unequivocally that background affects mating success. A male whose main color is predominant in the background will be at a sexual disadvantage to males whose main colors are rare in the background. Males which contrast more with their background are at an advantage to males which contrast less, but the contrast depends upon the particular background against which they are seen.

As shown by all experiments, males can be successful by contrasting with a given background by having spots which are a different size from the background as well as having colors which are uncommon in the background. Conspicuousness also applies to the predators, and there are five ways in which males manage to increase their conspicuousness to females, while minimizing their visibility to predators: (1) some of the colored spots are found in the dorsal fin, which is erected during courtship, but is partially or wholly folded at other times, hiding the colors. (2) Some of the colored spots may be small enough to be below the visual acuity or color detection threshold of the predators at their attack distance of 10–15 cm, but all spots can be seen by females (Endler 1978). (3) There is some nervous control over the size of some black spots; they are enlarged during courtship (Baerends et al. 1955) and are relatively small at other times. (4) Spots are often longer than high (Endler 1980); most foraging movements are along the long axis of

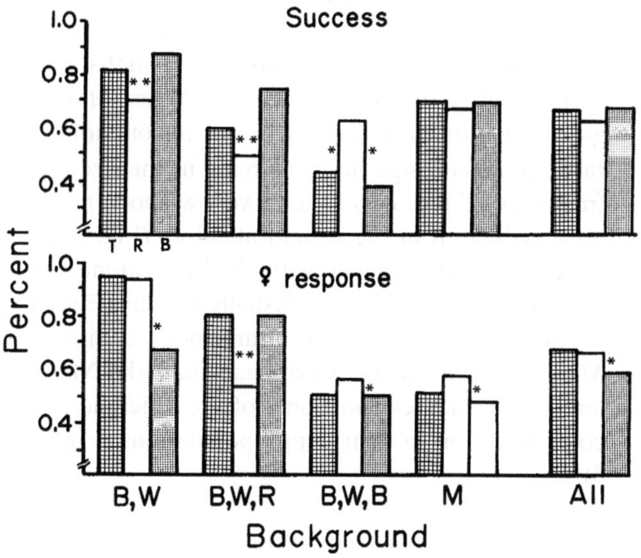

Fig. 5. Preliminary results of mate choice as a function of background color patterns. Vertical axes: percentage of presentations of two males to a single female which resulted in a successful copulation (above) or a significant female response (below). The backgrounds and number of trials (one trial per virgin female) are: B,W – black & white gravel (63). B,W,R – black, white & red gravel (45). B,W,B – black, white & blue gravel (48). M – multicolored gravel (75). All – all experiments pooled (231). The hatched bars (T) are for the male of a pair with more spots and the open bars (R) for males with more red or orange spots, and the shaded bars (B) for males with more blue spots. Asterisks indicate that adjacent bars are significantly different from each other. Note that having red on a non-red background, or blue on a non-blue background is better than having a color which is common in the background.

the body, whereas the sexual display consists of rapid vertical movements. During body movement, and aided by flicker fusion, the spots appear smaller during normal movement, and much larger during courtship. This predicts that it is more critical for patch height to match the background than patch length, and this is supported by the results of the greenhouse experiment (Endler 1980). (5) 'Private wavelengths', colors which can be seen by females, but not by predators.

If the most dangerous predators are incapable of seing certain wavelengths, then those colors should be more developed than the colors which the predators can detect (Endler 1978). This is borne out by a comparison between guppies living with *Rivulus hartii* alone (R), and with *R. hartii* and *Macrobrachium* (M or R + M). Prawns are relatively insensitive to long (red) wavelengths and more sensitive to short wavelengths. When *Macrobrachium* is present (and *R. hartii* uncommon to rare), there is a significant increase in the size and frequency of red, orange, and yellow spots, and a great reduction in the structural colors (Fig. 1, 2), relative to the localities with *R. hartii* alone but otherwise ecologically and visually similar (Endler 1978). The much greater abundance of diurnal *Macrobrachium* in the Caribbean than the Mainland fauna may explain some of the differences in color variation in their guppy populations (Fig. 2, 3).

Carotenoid versus structural colors and the basis for female choice

An interesting result of the female choice experiments (Fig. 5) is that red is apparently at an advantage in the initial stages of courtship (female response), while blue is better once the female's attention is gained, correcting for the background effect. This may result from the differing properties of structural and carotenoid colors. The brightness of carotenoid colors (red, orange, yellow) depends upon diet while structural colors are independent of diet (Fox & Vevers 1960, Rothschild 1975, Davies 1976). If fish are fed a carotenoid-free (or carotenoid-poor) diet, then these colors fade over a few weeks, and if carotenoids are provided again, the colors become bright and intense.

Since they depend upon diet, carotenoid colors can be a direct indication of food finding ability (Endler 1980). In undisturbed habitats, guppies live in oligotrophic streams with clean gravel bottoms in montane forest, and there is a premium on food, especially food containing carotenoids such as algae. Natural populations in Trinidad eat mainly benthic algae and invertebrates (Dussault 1980); algae are a major source of carotenoids. Any male which consumes more carotenoids during foraging will have brighter carotenoid colors. Thus male brightness, at least with respect to carotenoid colors, is a direct indicator of feeding success. This makes no assumptions about males specifically seeking out carotenoid-rich foods, but merely that a male with a faster ingestion rate per day will incidentally pick up more carotenoids and become brighter. Carotenoid colors vary in brightness in natural populations, especially in streams through the densest forest. This has a profound implication for sexual selection. If carotenoids were used as indicators of food finding ability, then females may prefer males with brighter, more, or larger carotenoid patches.

If females simply select on the basis of brightness, or absolute contrast with the background (Fig. 5), then males with bright structural colors will be favored by females also. Structural colors are based upon refraction and reflection through guanine crystals and are independent of feeding success. Thus, in some cases, especially in areas of low food availability and low predation, males with structural colors will be favored over those with carotenoid colors. But this cannot go too far because then the correlation between male brightness and male fitness will decline to zero. As long as there is predation, females which waste time choosing males (courtship is a predation vulnerable time) and obtain nothing in return, will be at a disadvantage, and the frequency of discriminating females will decline until an equilibrium is reached. The system is analogous to a Batesian-Mullerian mimicry system (many spots of all classes) in that the structural colors are false advertising for male fitness (Endler 1980); the brightness of structural colors is frequently greater than the carotenoids and diet independent. Thus, in a sense the carotenoids are the model, and the structurals the mimics.

The mimicry analogy predicts that we should never find guppies with only structural colors (mimics), but we should find guppies with only carotenoids (models) or with both color classes. In more than 10^5 guppies scored, I have never seen guppies with only structural colors, but I have seen some with only carotenoids and most with both. In addition to the analogue of the equilibrium between models and mimics, there is a second prediction. It would be good for females to favor males with more carotenoid colors, but a genetic tendency to select for structural colors would be disadvantageous. This is supported by the mate choice experiments, which indicate that males with more carotenoids are more successful in the early stages of courtship (Fig. 5).

The second stage in courtship, copulation, appears to be dependent upon structural as well as carotenoid colors (Fig. 5). It should be emphasized that it is the subset of males which have been successful in gaining the female's attention, and have more structural colors, that are more successful in copulation, and these do not necessarily have more structurals than the general population. There are two possibilities for structural colors functioning in female choice: (1) Structurals may be more efficient in communication during sexual display in the low light intensity in undisturbed streams; structural colors reflect more light than carotenoids, and thus may be more 'stimulating'. (2) Because structural colors reflect the highest proportion of incident light at a particular angle, females see a flashing of light from the blue and silver spots during the male's vibrating display. The flash frequency is a direct and efficient estimate of the speed and duration of vibration. It is possible that healthier males can vibrate, hence flash, more rapidly and longer than less healthy males. Thus the flicker from structural colors may also be an indicator of male fitness. This is currently being investigated.

Although incomplete, the data on sexual selection and female choice clearly show that female choice depends primarily upon absolute criteria (contrast with a particular background) rather than relative criteria (contrast between males), and that some aspects of male color patterns may be direct indicators of male fitness. The fitness-indicating hypothesis is reasonable.

The Fisherian models of sexual selection (Lande 1980, 1982, Kirkpatrick 1982) are possible alternative hypotheses to explain sexual selection in guppies. The basic model assumes that there is genetic variation in a male character, and genetic variation among females in their willingness to accept a male with any given character value. This sets up a correlation between male character and female choice criteria which can evolve rapidly, by a correlated response among the male and female characters. Males with more extreme values may be favored by females, which in turn favors females which choose more extreme males, and the system may 'runaway' to extreme values of male characters and female choice criteria. The model makes no assumptions about the fitness of the male character, and the male character may even be maladaptive. Females choosing less viable males are as likely to evolve as females choosing more viable males.

The main problem with the Fisherian models is the problem of what starts the process, and why the female should use a particular criterion to begin with. Once there is a correlation between the male character and female criterion, then the system will rapidly evolve. In guppies the system could have been started by chance and by chance lead to selection by females of males on the basis of absolute contrast with the background. Alternatively, absolute contrast may have been initially favored because such males were simply easier to see in the dark streams. But these models do not explain why carotenoid colors should be favored over structural colors in courtship, or the observed carotenoid/structural color ratios in natural populations. This is easier to explain on the basis of direct fitness indication by males through their carotenoid colors, with the structural color mimics tagging along. Of course a hybrid model is possible in which fitness indication is allowed in a Fisherian model, but computer simulations (Endler unpublished) indicate that the effects of fitness indication and predation overwhelm the effects of the Fisher runaway process. When fitness indicating genes (such as those controlling carotenoid spots) are present, then the females choosing 'better' males win at the expense of those choosing 'poorer' (duller) males. Finally, all of the variation in guppy color patterns can be explained by the fitness indicating model in the absence of the Fisher runaway process. Therefore, the Fisherian model is either superfluous, or in the absence of fitness indication does not explain as much of the data as does the fitness indication hypothesis.

Summary of guppy data

The evidence strongly supports the hypothesis that the extensive color pattern polymorphism in *Poecilia reticulata* is affected by predation and sexual selection, interacting with background color patterns. These factors interact in predictable ways. The color patterns in a particular place represent a balance between sexual selection, which tends to increase conspicuousness against the local background, and visually mediated predation, which tends to decrease conspicuousness. The balance is very different between the three classes of colors. Melanin spots (black) are voluntarily increased in size during courtship, but reduced in size and intensity at other times. Carotenoid spots (red, orange, yellow) genetically decrease in size and frequency with increased predation. They are often longer than high, making them more conspicuous when movement is vertical than when it is horizontal, and are difficult for arthropod predators to see. Carotenoid colors may also be a direct indication of how well fed a male is and has been for the past few weeks. Structural colors (blue, iridescent, silver, bronze) genetically decrease in frequency with increased predation, are sometimes found in the folding dorsal fin, may be longer than wide, and may be direct indicators of male physical fitness during their flicker in courtship. Because these three classes of colors differ in properties, functions and mode of natural selection, they cannot be treated as equivalent color pattern elements in any evolutionary study.

Color pattern polymorphism in other species

Only three other species have been examined in any detail: *Phalloceros caudimaculatus*, *Xiphophorus maculatus* and *X. variatus*. *Phalloceros caudimaculatus* is ecologically similar to *Poecilia reticulata*, but lives in southeastern South America, chiefly southern Brazil (Rosen & Bailey 1963, Endler 1982). Like the guppy, it is the only member of its family through most of its natural range, and tends to be found in small creeks and streams, though they tend to be sandier and slower moving than guppy streams. As with guppy populations, it is possible to rank stream segments for visual predation intensity by the number and kinds of predators present. Unlike guppies, there is very little sexual dimorphism, and the color pattern polymorphism is much simpler (Fig. 1d,e). *P. caudimaculatus* varies in the number and size of black spots, and in the frequency of colored dorsal and caudal fins. The dorsal and caudal fins are either clear or various shades of red or orange. Unfortunately, the genetics of these characters is unknown. In spite of these limitations, there is a highly significant correlation between rank predation intensity and the color pattern characteristics. As predation intensity increases, the number and size of black spots decreases and the frequency of fish with colored fins decreases (Endler 1982). The results exactly parallel the results for guppies at the other end of South America, living with entirely different predators.

A comparison between streams with weak or no predation (the latter not found with guppies) and streams with only arthropod predators (freshwater prawns, dytiscid beetles, belostomatid beetles, large odonate larvae) shows that black spots are large and the proportion of colored fins is high in places with no or weak predators, but where there are only arthropod predators the black spots are tiny and the frequency of colored fins is high. By contrast, whenever there are intermediate or dangerous fish predators present, the black spots are tiny and the frequency of colored fins is low. As with guppies living with prawns, the only cryptic color patterns are those which the predators can see (Endler 1982). There appears to be good reason to believe that the relationships between color pattern characteristics, predation and sexual selection is a general one, and not just limited to guppies.

Xiphophorus are morphologically, geographically, and ecologically quite different from *Poecilia* and *Phalloceros* (Rosen & Bailey 1963, Miller 1966, Rosen 1979). In contrast to guppies and *P. caudimaculatus*, *Xiphophorus* species live in a greater variety of habitats and elevations, and usually with a variety of other Poeciliids, and sometimes with congeners. Thus species recognition may limit or affect the development of their color patterns. The background color patterns may also be very different, because *Xiphophorus* live in streams, lakes, and ponds with silty and muddy beds, as well as places with dense growths of aquatic and semi-aquatic vegetation – these habitats are very rare for guppies and *P. caudimaculatus*.

Xiphophorus maculatus and *X. variatus* have a similar but simpler polymorphism than guppies (Kallman 1975), and something is known about social behavior in relation to color pattern (Borowsky & Kallman 1976, Borowsky & Khouri 1976, Borowsky 1978a,b), but little is known about their predators or ecology. This is complicated by the fact that most of their natural habitat is extensively disturbed by man. To make matters more complicated, the differences between color pattern morphs are correlated with differences in body size, and it is body size which determines mating success (Borowsky 1981). Thus, in *Xiphophorus* we do not know whether there is natural selection for color patterns or for body size and growth rate; one or the other characters may be 'hitch-hiking' on the gene frequency changes of the other (Thomson 1977). This is because genes for some color pattern elements are closely linked with growth and maturation genes (Kallman & Borkoski 1978). There are also some fascinating correlations between some of the color pattern element genes and apparent resistance to environmental stress (Borowsky 1978a, 1981); whether or not these are a result of closely linked genes or pleiotropic effects, or are primary results of the color pattern genes, this is obviously worth further research.

The *Xiphophorus* results are in strong contrast to those for guppies and *P. caudimaculatus*. Although there is a suggestion that size may influence mating success in guppies (Kodric-Brown personal communication 1976), there is little body size variation within natural guppy populations (see Fig. 1, and color plates in Haskins et al. 1961, Endler 1978); most variation is between populations with differing predation intensities (Liley & Seghers 1975, Endler 1980). There is no correlation between color morphs and body size (Endler unpublished), and Farr (1980a) and I found that courtship display rate and color pattern are the major determinants of courtship success in guppies. In guppies and *P. caudimaculatus*, predation and sexual selection are the main determinants of color patterns, while in *Xiphophorus* physiological effects, possibly at closely linked loci, seem more important.

It would be of very great interest to investigate the predator communities of *Xiphophorus* streams for any effects on the color patterns. There is the barest suggestion that predation is important. There is a positive correlation between adult male size and the proportion of *crescent (C)* alleles in natural populations, and *C* males tend to be larger than the allele *cut-crescent (Ct*, Borowsky 1978). This is complicated by the fact that *C* males tend to be more variable than *Ct* males. In addition, the spot size of *Ct* is smaller than *C*. In both guppies and *P. caudimaculatus* male size is consistently and significantly smaller in high predation areas than in low predation areas (Liley & Seghers 1975, Endler 1980, 1982). The positive correlation between mean male size and proportion of males with *C* means that there is a positive correlation between male size and mean spot size among *Xiphophorus* localities. This is entirely consistent with geographic variation in predation intensity: as predation intensity increases, males become smaller and the proportion of males with smaller spots (*Ct*) increases. This is worth further investigation.

Conclusions

Although a casual glance at the color patterns of poeciliids may make one think that the color patterns are random, detailed study of South American species has shown that the color patterns are very important in (a) avoiding diurnal visually hunting predators, (b) mating success and (c) species recognition. Data from Central American *Xiphophorus* indicate that some color pattern elements may be closely linked to physiology-affecting loci, which further affect the variation in color patterns.

Using the theory of color pattern and background matching (crypsis) and conspicuousness (Endler 1978) it is possible to predict in detail the kinds of color patterns that will best serve various purposes on particular backgrounds. These tests have been quite successful in poeciliids. It is important to note that different color patterns have different degrees of conspicuousness on different backgrounds, and their degrees of crypsis may be different to predators and mates with different visual abilities.

Acknowledgements

I am most grateful for good discussions and comments on an earlier version of this paper by Jeff Baylis, Ric Charnov, Tom Zaret and Mark Kirkpatrick. I am pleased to acknowledge financial support from the National Geographic Society, the Center for Field Research, and the U.S. National Science Foundation grants BMS 75-11903, DEB 78-11200, DEB 79-27021 and DEB 82-00295.

References cited

Baerends, G.P., R. Brouwer & H.T. Waterbolk. 1955. Ethological studies on *Lebistes reticulatus* (Peters). I. An analysis of the male courtship pattern. Behaviour 8: 249–335.

Balsano, J.S., K. Kucharski, R.J. Randle, E.M. Rasch & P.J. Monaco. 1981. Reduction of competition between bisexual and unisexual fishes of *Poecilia* in northeastern Mexico. Env. Biol. Fish. 6: 39–48.

Balon, E.K. 1975. Reproductive guilds of fishes: a proposal and definition. J. Fish. Res. Board Can. 32: 821–864.

Balon, E.K. 1983. Patterns in the evolution of reproductive styles in fishes. *In:* G.F. Potts & R.J. Wootton (ed.) Fish Reproduction: Strategies and Tactics, Academic Press, London. (In print).

Borowsky, R.L. 1973. Relative size and development of fin coloration in *Xiphophorus variatus*. Physiol. Zool. 46: 22–28.

Borowsky, R.L. 1978a. The tailspot polymorphism of *Xiphophorus* (Pisces: Poeciliidae). Evolution 32: 886–893.

Borowsky, R.L. 1978b. Social inhibition of maturation in natural populations of *Xiphophorus variatus* (Pisces: Poeciliidae). Science 201: 933–935.

Borowsky, R.L. 1981. Tailspots of *Xiphophorus* and the evolution of conspicuous polymorphisms. Evolution 35: 345–358.

Borowsky, R.L. & K.D. Kallman. 1976. Patterns of mating in natural populations of *Xiphophorus* (Pisces: Poeciliidae). I. *X. maculatus* from Belize and Mexico. Evolution 30: 693–706.

Borowsky, R.L. & J. Khouri. 1976. Patterns of mating in natural populations of *Xiphophorus* (Pisces: Poeciliidae). II. *X. variatus* from Tamaulipas, Mexico. Copeia 1976: 727–734.

Byron, E.R. 1981. Metabolic stimulation by light in a pigmented freshwater invertebrate. Proc. Nat. Acad. Sci. U.S.A. 78: 1765–1767.

Clark, E. & L.R. Aronson. 1951. Sexual behavior in the guppy, *Lebistes reticulatus*. Zoologica 36: 49–66.

Constantz, G.D. 1975. Behavioral ecology of mating in the male gila topminnow, *Poeciliopsis occidentalis* (Cyprinodontiformes: Poeciliidae). Ecology 56: 966–973.

Cott, H.B. 1940. Adaptive coloration in animals. Methuen and Co., London. 508 pp.

Davies, B.H. 1976. Carotenoids. pp. 38–165. *In*: T.W. Goodwin (ed.) Chemistry and Biochemistry of Plant Pigments, Academic Press, New York.

Dussault, G.V. 1980. Feeding behavior in the guppy, *Poecilia reticulata* (Pisces: Poeciliidae). M.S. Thesis, McGill University, Montreal. 71 pp.

Endler, J.A. 1978. A predator's view of animal color patterns. Evol. Biol. 11: 319–364.

Endler, J.A. 1980. Natural selection on color patterns in *Poecilia reticulata*. Evolution 34: 76–91.

Endler, J.A. 1982. Convergent and divergent effects of natural selection on color patterns in two fish faunas. Evolution 36: 178–188.

Farr, J.A. 1976. Social facilitation of male sexual behavior, intrasexual competition, and sexual selection in the guppy, *Poecilia reticulata* (Pisces, Poeciliidae). Evolution 30: 707–717.

Farr, J.A. 1977. Male rarity or novelty, female choice behavior, and sexual selection in the guppy, *Poecilia reticulata* Peters (Pisces, Poeciliidae). Evolution 31: 162–168.

Farr, J.A. 1980a. Social behavior patterns as determinants of reproductive success in the guppy, *Poecilia reticulata* Peters (Pisces, Poeciliidae). An experimental study of the effects of intermale competition, female choice, and sexual selection. Behaviour 74: 38–91.

Farr, J.A. 1980b. The effects of sexual experience and female receptivity on courtship-rape decisions in male guppies, *Poecila reticulata* (Pisces: Poeciliidae). Anim. Behav. 28: 1195–1201.

Farr, J.A. & W.F. Herrnkind. 1974. A quantitative analysis of social interaction of the guppy, *Poecilia reticulata* (Pisces, Poeciliidae), as a function of population density. Anim. Behav. 22: 582–591.

Fisher, R.A. 1930. The evolution of dominance in certain polymorphic species. Amer. Nat. 64: 385–406.

Fox, H.M. & G. Vevers. 1960. The nature of animal colors. Sidgewick & Jackson, London.

Gandolfi, G. 1971. Sexual selection in relation to social status of males in *Poecilia reticulata* (Teleostei, Poeciliidae). Boll. Zool.

38: 35–48.

Ghiselin, M.T. 1974. The economy of nature and the evolution of sex. Univ. California Press, Berkeley. 346 pp.

Gorlick, D.L. Dominance hierarchies and factors influencing dominance in the guppy, *Poecilia reticulata* Peters. Anim. Behav. 24: 336–346.

Greene, R.J., Jr. 1972. Female preferential selection for males in *Lebistes reticulatus*. Undergraduate Biology thesis, Univ. Utah, Salt Lake City.

Haskins, C.P. & E.F. Haskins. 1949. The role of sexual selection as an isolating mechanism in three species of Poeciliid fishes. Evolution 3: 160–169.

Haskins, C.P. & E.F. Haskins. 1950. Factors governing sexual selection as an isolating mechanism in the Poeciliid fish *Lebistes reticulatus*. Proc. Nat. Acad. Sci. U.S.A. 36: 464–476.

Haskins, C.P., E.F. Haskins, J.J.A. McLaughlin & R.E. Hewitt. 1961. Polymorphism and population structure in *Lebistes reticulatus*, a population study. pp. 320–395. *In*: W.F. Blair (ed.) Vertebrate Speciation, Univ. Texas Press, Austin.

Hildemann, W.H. & E.D. Wagner. 1954. Intraspecific sperm competition in *Lebistes*. Amer. Nat. 88: 87–91.

Hynes, H.B.N. 1970. The ecology of running waters. Univ. Liverpool Press, Liverpool 555 pp.

Jacobs, K. 1971. Livebearing aquarium fishes. MacMillan, New York. 461 pp.

Kallman, K.D. 1975. The platyfish, *Xiphophorus maculatus*. pp. 81–132. *In*: R.C. King (ed.) Handbook of Genetics, vol. 4, Vertebrates of Genetic Interest, Plenum Press, New York.

Kallman, K.D. & V. Borkoski. 1978. A sex-linked gene controlling the onset of sexual maturity in female and male platyfish (*Xiphophorus maculatus*), fecundity in females and adult size in males. Genetics 89: 79–119.

Keenleyside, M.H.A. 1979. Diversity and adaptation in fish behavior. Springer Verlag, New York. 208 pp.

Kennedy, C.E.J. 1979. Factors influencing the sexual behavior of the guppy, *Poecilia reticulata*. Ph.D. Thesis, Univ. Leicester, Leicester, England. 197 pp.

Kirkpatrick, M. 1982. Sexual selection and the evolution of female choice. Evolution 36: 1–12.

Lande, R. 1980. Sexual dimorphism, sexual selection, and adaptation in polygenic characters. Evolution 34: 292–305.

Lande, R. 1981. Models of speciation by sexual selection on polygenic traits. Proc. Nat. Acad. Sci. U.S.A. 79: 3721–3725.

Lewontin, R.C. 1974. The genetic basis of evolutionary change. Columbia Univ Press, New York. 346 pp.

Liley, R.N. 1966. Ethological isolating mechanisms in four sympatric species of Poeciliid fishes. Behaviour Supplement 13: 1–197.

Liley, R.N. & B.H. Seghers. 1975. Factors affecting the morphology and behavior of guppies in Trinidad. pp. 92–118. *In*: G.P. Baerends, C. Beer & A. Manning (ed.) Function and Evolution in Behavior, Oxford Univ. Press, Oxford.

Martin, F.D. & M.F. Hengstebeck. 1981. Eye color and aggression in juvenile guppies, *Poecilia reticulata* (Pisces: Poeciliidae). Anim. Behav. 29: 325–331.

Martin, R.G. 1977. Density-dependent advantage in melanistic male *Gambusia*. Florida Sci. 40: 393–400.

Miller, R.R. 1966. Geographical distribution of Central American freshwater fishes. Copeia 1966: 773–802.

Peters, G. 1964. Vergleichende Untersuchungen an drei Subspecies von *Xiphophorus helleri* (Haeckel) (Pisces). Z. Zool. Syst. Evol. 2: 185–271.

Reznick, D.N. 1980. Life history evolution in the guppy (*Poecilia reticulata*). Ph.D. Thesis, Univ. Pennsylvania, Philadelphia. 224 pp.

Reznick, D.N. & J.A. Endler. 1982. The impact of predation on life history evolution in Trinidadian guppies (*Poecilia reticulata*) Evolution 36: 160–177.

Rosen, D.E. 1979. Fishes from the uplands and intermontane basins of Guatemala: revisionary studies and comparative biogeography. Bull. Amer. Mus. Nat. Hist. 162: 268–375.

Rosen, D.E. & R.M. Bailey. 1963. The Poeciliid fishes (Cyprinodontiformes), their structure, zoogeography, and systematics. Bull. Amer. Mus. Nat. Hist. 126: 1–176.

Rosen, D.E. & M. Gordon. 1953. Functional anatomy and evolution of male genitalia in Poeciliid fishes. Zoologica 38: 1–48.

Rosen, D.E. & K.D. Kallman. 1969. A new fish of the genus *Xiphophorus* from Guatemala, with remarks on the taxonomy of endemic forms. Amer. Mus. Novit. 2379: 1–29.

Rosen, D.E. & A. Tucker. 1961. Evolution of secondary sexual characters and sexual behavior patterns in a family of viviparous fishes (Cyprinodontiformes: Poeciliidae). Copeia 1961: 201–212.

Rothschild, M. 1975. Remarks on carotenoids in the evolution of signals. pp. 20–47. *In*: L.E. Gilbert & P.H. Raven (ed.) Coevolution of Animals and Plants, Univ. Texas Press, Austin.

Schreibman, M.P. & K.D. Kallman. 1978. The genetic control of sexual maturation in the teleost *Xiphophorus maculatus* (Poeciliidae); a review. Ann Biol. Anim. Biochem. Biophys. 18: 957–962.

Seghers, B.H. 1973. An analysis of geographic variation in the antipredator adaptations of the guppy, *Poecilia reticulata*. Ph.D. Thesis, University of British Columbia, Vancouver. 273 pp.

Thomson, G. 1977. The effect of a selected locus on linked neutral loci. Genetics 85: 753–788.

Warburton, B., C. Hubbs & D.W. Hagen. 1957. Reproductive behavior of *Gambusia heterochir*. Copeia 1957: 299–300.

Whitton, B.A. (ed.). 1975. River ecology. Univ. California Press, Berkeley. 725 pp.

Wickler, W. 1968. Mimicry in plants and animals. Weidenfield & Nicholson World University Library, London. 255 pp.

Winge, O. 1937. Succession of broods in *Lebistes*. Nature 140: 467.

Wourms, J.P. 1981. Viviparity: the maternal-fetal relationship in fishes. Amer. Zool. 21: 473–515.

Yamamoto, T. 1975. The medaka, *Oryzias latipes*, and the guppy, *Lebistes reticulatus*. pp. 133–149. *In*: R.C. King (ed.) Handbook of Genetics, vol. 4, Vertebrates of Genetic Interest, Plenum Press, New York.

Zander, C.K. 1965. Die Geschlechtsbestimmung bei *Xiphophorus montezumae cortezi* Rosen (Pisces). Z. Vererbungslehre 96: 128–141.

Originally published in Env. Biol. Fish. 9: 173–190

Gondwana and neotropical galaxioid fish biogeography

Hugo Campos
Instituto de Zoologia, Universidad Austral de Chile, Casilla 567, Valdivia, Chile

Keywords: Anadromous fishes, Dispersal, Panaustral distribution, Phylogeny, Reproductive ecology, Salmoniformes, Systematics, Vicariance

Synopsis

The galaxioids, Southern Hemisphere salmoniforms (i.e. Galaxiidae, Aplochitonidae, Retropinnidae and Prototroctidae), are found in Australia, New Zealand, New Caledonia, South America and South Africa, and include 4 families, 11 genera, and about 45 species. Biogeographers have disagreed on the interpretation of this distributional pattern, shared by many Southern Hemisphere taxa. Both ecological and systematic data are important in resolving what has become a controversial topic among students of biogeography.

Data on the reproductive ecology of land-locked populations show that the estuarine phase in galaxioids is not a true marine dispersive phase. Ecological and morphological data from sympatric, but possibly reproductively isolated, lacustrine populations, suggest that the controversial species *Galaxias alpinus* has taxonomic validity. The theory can be separated into ecological (i.e. via population dispersal) versus historical (i.e. allopatric speciation or vicariant events). According to the former, Australia was the center of origin, with subsequent dispersion by successive invasions via New Zealand through the East Australian Ocean Current. Finally, as a result of west wind drift, the species reached South America and eventually South Africa. The principal emphasis for this point of view, therefore, has been to attempt to explain marine dispersal through the presence of the juvenile phase (whitebait) in several species of diadromous *Galaxias*. In contrast, the historical theory supposes a Gondwana pattern of distribution. This is supported by studies on dependency on fresh water of most galaxioids (e.g. *Galaxias maculatus*) and a broad distribution (South America, New Zealand and Australia). We suggest that the dispersion of *G. maculatus* occurred after the actual pattern of galaxioid distribution. The best explanation for the disjunct distribution of these fishes is a historical explanation, based on the disruption of the continent of Gondwana.

Introduction

The galaxioid fishes (Galaxiidae, Aplochitonidae, Retropinnidae and Prototroctidae), salmonoids of the southern hemisphere, have been a controversial topic among students of biogeography because of their disjoint distribution among the continents and islands that surround the Antarctic. Their distributional pattern is similar to that documented for other southern hemisphere plants and animals. Some authors have found an explanation for these patterns in the disruption of the hypothetical continent of Gondwana, others have assumed dispersion mechanisms, and still others the crossing of oceanic barriers to account for the families' disjunct distributions. This contribution considers: (1) the systematic and taxonomic studies which provide the basis for the validity of the taxa being discussed and

Thomas M. Zaret (ed.), Evolutionary ecology of neotropical freshwater fishes. ISBN 90 6193 823 6

their phylogenetic deductions; (2) the life cycle and ecology of the species, and their capacity to disperse; and (3) the zoogeographic studies, cladistic phylogenetic analyses, and their application to the geological history of the galaxioids.

Systematics

The families of the order Salmoniformes (Greenwood et al. 1966) present a panboreal and panaustral distribution (Fig. 1). The Galaxiidae, Aplochitonidae, Retropinnidae, Prototroctidae and Salangidae families of the southern hemisphere are related to salmonoids of the northern hemisphere. The classification of these families was discussed first by Günther (1880) and recently by Rosen (1974). First knowledge of the southern families suggested that they had affinities with the north. Regan (1905) suggested a similarity of Esocidae with Galaxiidae and Aplochitonidae in some characteristics, but important differences in others. Moreover, Regan (1913) noted that the Galaxiidae and Aplochitonidae were more specialized than the Osmeridae (smelts) of the northern sea. Chapman (1944) also suggested that galaxiid and esocoid fishes were closely related. Gosline (1960) maintained that 'the Aplochitonidae, Retropinnidae and Galaxiidae are derivatives of a protoosmerid stock . . .' to which Weitzmann (1967) agreed. McDowall (1969) established that the relationships of these families were close to the Osmeroidei. Nelson

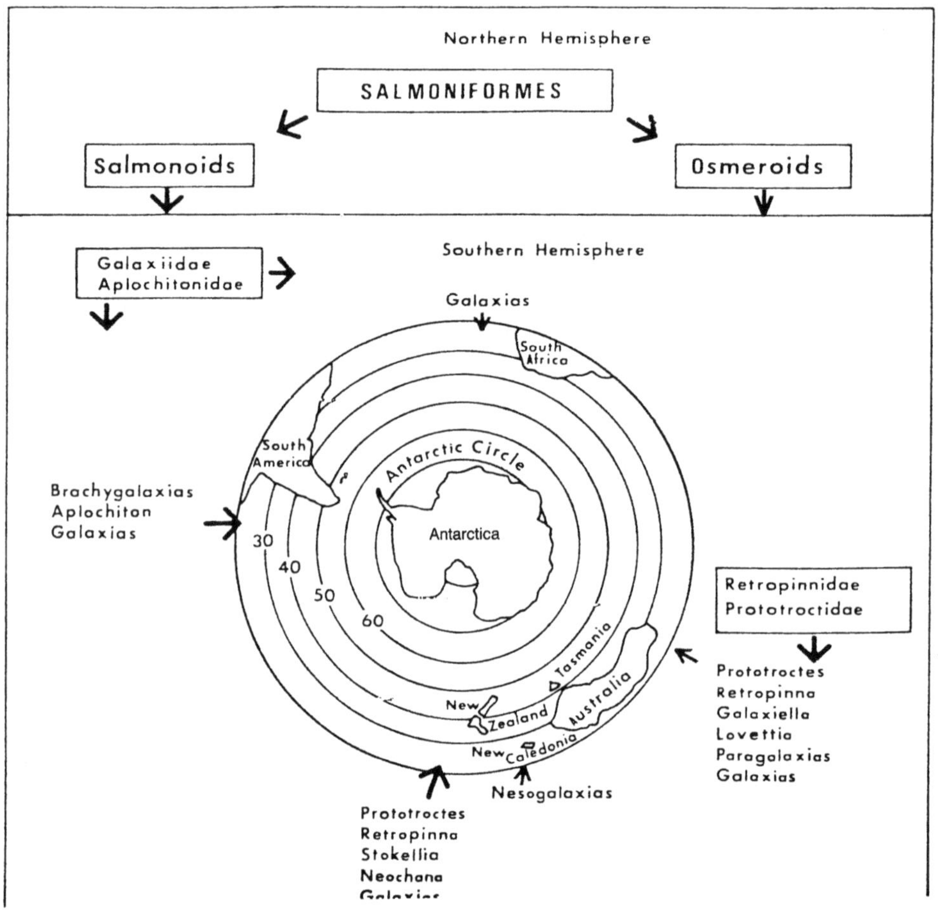

Fig. 1. Distribution of Galaxioids in the Southern Hemisphere and affinity to salmoniforms from the Northern Hemisphere.

(1972), through studies on the cephalic and pitlines sensorial channels, established that Chapman's (1944) proposed relationships between galaxiids and esocoids did not hold. Finally, Rosen (1974) presented a possible phylogeny of the Salmoniformes relying heavily on the anatomy of the hypobranchial apparatus which can be used to establish levels of relationships among the Salmoniformes. Rosen (1974) considers that the esocoids are the sister group of the argentinoids, galaxioids (galaxiids and aplochitonids), salmonoids and osmeroids (osmerids, plecoglossids, salangids, retropinnids and prototroctids). Galaxioids and salmonoids constitute a monophyletic group; argentinoids appear more primitive than galaxioids, salmonoids or osmeroids.

In attempting to determine whether these families constitute a monophyletic lineage, the most problematic family has been Salangidae which is found only in the Orient (China, Japan and Korea). Gosline (1960) suggests that Salangidae are early derivatives of the osmerid-plecoglossid line, and that their similarity to the aplochitonid *Lovettia* is explainable as neotenic development. Greenwood et al. (1966) include Salangidae in the sub-order Galaxioidei together with the other southern families. Weitzmann (1967) thinks that the Salangidae should be considered more closely related to the northern Osmeridae. McDowall (1969) provides a list of osteologic characters that would unify all the galaxioid fishes (Galaxiidae, Aplochitonidae, Retropinnidae and Prototroctidae). He concludes that the Galaxiidae and Retropinnidae are natural phyletic units and the family Aplochitonidae is divided into Prototroctidae and Aplochitonidae. Nelson (1972) clearly establishes that retropinnids and prototroctids are closely related, suggesting a subdivision of the Galaxiidae into a sub-family Retropinninae (*Prototroctes*, *Retropinna* and *Stokellia*) and a sub-family Galaxiinae (*Aplochiton*, *Brachygalaxias*, *Galaxias*, *Lovettia*, *Neochana*, *Nesogalaxias* and apparently *Saxilaga* and *Paragalaxias*). Rosen (1974) concludes that the galaxiids and aplochitonids are related to salmonoids and that retropinnids and prototroctids are related to osmerids, salangids and plecoglossids. The anatomical structures used in his phylogenetic analysis suggest to him a mosaic evolution in the Salmoniformes and that only the hypobranchial apparatus can be used to establish valid criteria in the whole group. Despite the fact that the relationships among the galaxioid families are still being discussed, it would seem that the Salangidae are not close. I consider as southern salmoniforms the families Galaxiidae, Aplochitonidae, Retropinnidae and Prototroctidae.

The family Galaxiidae, first established by Müller (1844), has unique characteristics: they are not scaled, the dorsal fin is displaced backwards above the anal fin and they lack an adipose fin. The family includes six genera recognized by McDowall (1969) and McDowall & Frankenburg (1981): *Galaxias*, *Brachygalaxias*, *Neochana*, *Nesogalaxias*, *Paragalaxias* and *Galaxiella*. The validity of other genera such as *Mesites* Jenyns 1842, *Austrocobitis* Ogilby 1899, *Agalaxias* Scott 1966, *Querigalaxias* Whitley 1935 and *Lyragalaxias* Whitley 1935 has been rejected by McDowall (1969). The family presently contains approximately 45 species, and is widely distributed in southern America, South Africa Australia, Tasmania and New Zealand. The genus *Galaxias* Cuvier 1817 is the most abundant in number of species (24) and is present through the family distribution. The genus *Brachygalaxias* Eigenmann 1928 includes only one species (McDowall 1973a), restricted to Chile (previously though to exist in Tasmania, Scott 1942). The genus differs from the other genera because it has a primitive caudal skeleton where at least two centra are involved in the urostyle (Greenwood et al. 1966) and its maxillary alveolar process is elongated, totally excluding the maxilla from the gape (McDowall 1969). The genus *Neochana* Günther 1967 with three species in New Zealand has its pelvic fins reduced or absent. The genus *Nesogalaxias* Whitley 1935, with one species in New Caledonia, according to McDowall (1968) differs 'in the absence of pleural ribs from those vertebrae behind the pelvic fins'. The genus *Paragalaxias* Scott 1935 has four species in Tasmania, and differs from other galaxiids in the origin of the dorsal fin above the pelvic fins.

Galaxiidae

The species of the family Galaxiidae constitute a monophyletic group (McDowall 1969). Nevertheless, phylogenetic relationships at the species level are still unresolved. McDowall (1970) suggests phylogenetic relationships among species in New Zealand, proposing five groups of which four are *Galaxias* and one unites the three species of *Neochana*. Our study (Campos 1979) of the species of South America, South Africa and New Zealand, using numerical taxonomy and principal components analysis, supported the groups proposed by McDowall (1970). The analysis demonstrated a weak relationship within the species of South America. *Brachygalaxias bullocki* of South America was grouped in the same cluster with *Galaxias zebratus* of South Africa. The generic position of *zebratus* has been discussed, attributing it to the genera *Paragalaxias* (Stokell 1950), *Agalaxias* (Scott 1966), and maintained in the genus *Galaxias* by a majority of authors (Regan 1905, Barnard 1943, McDowall 1973b, 1973a). On the other hand the genus *Brachygalaxias* has been attributed to *Galaxias pusillus* of Victoria, Australia (Stokell 1954, Scott 1942, Frankenberg 1974) and with three subspecies (Scott 1971), *Brachygalaxias pusillus tasmaniensis*, of Tasmania, *B. p. flindersiensis* of the Flinders Island and the species *Brachygalaxias nigrostriatus* (Shipway). McDowall (1973a) concludes that *nigrostriatus* and *pusillus* belong to *Galaxias* rather than *Brachygalaxias* though later on McDowall (1978) erects *Galaxiella* where he includes such species as *Galaxiella pusilla*, *G. nigrostriata* and new species *G. munda*. Another result of our analysis was to group *Galaxias globiceps* of Chile with *Neochana burrowsius* of New Zealand. The similarity between these two species was already observed by Scott (1936) who would have placed them together in the genus *Saxilaga*. *Galaxias platei* of Chile does not appear related to any single species but to the whole *maculatus* group.

The species *Galaxias maculatus* deserves special attention due to its wide distribution. The species was first described, by Jenyns 1842, as *Mesites attenuatus* from New Zealand and *Mesites maculatus* from Chile. They were later included in the genus *Galaxias* as *G. attenuatus* of New Zealand, and *G. maculatus* of South America and the Falkland Islands, but also mentioned for Australia and Chile (Cuvier & Valenciennes 1846, Günther 1866, Regan 1905, Eigenmann 1928). The conspecificity of both species was first noted by Stokell (1949, 1966) who concluded that only one species, *Galaxias maculatus*, is present in Australia, New Zealand and South America. Stokell (1966) proposed three sub-species: *Galaxias maculatus maculatus* (Jenyns) of South America, *G. m. attenuatus* (Jenyns) of New Zealand and *G. m. ignotus* of Australia. Scott (1968) disagreed and considered *G. maculatus* and *G. attenuatus* valid. McDowall (1967, 1970, 1971b), after studying material from New Zealand and South America, followed Stokell and assumed only one species, *Galaxias maculatus*.

In our cytogenetic studies (Campos 1972), the karyotype of *Galaxias maculatus* of Australia and Chile was compared with that of *Galaxias platei* and *Brachygalaxias bullocki*. The karyotypes of *G. maculatus* of Chile and Australia have the same number of chromosomes ($2n = 22$), with 8 metacentric, 12 submetacentric and 2 teleocentric. *Galaxias platei* is different from *G. maculatus* with $2n = 30$ chromosomes and *Brachygalaxias bullocki* with $2n = 38$ chromosomes. Subsequently, Merriles (1975) described the karyotype of *G. maculatus* of New Zealand as coinciding with the number of chromosomes ($2n = 22$) of those of Chile and Australia, but differing in the composition of the karyotype with 8 metacentric, 10 submetacentric and 4 subteleocentric. These results led us to postulate that in Chile and Australia the population studied are conspecific, but the New Zealand relationship is not clear. The morphological similarity between the populations of Chile and New Zealand is, however, very high (Campos 1972).

Another problem related to *Galaxias maculatus* is the species *G. alpinus* described for Chile by Jenyns (1842) as *Mesites alpinus*. Cuvier & Valenciennes (1846) recognized the species as *Galaxias alpinus* when creating the genus *Galaxias*, as did Regan (1905). Stokell (1966) proposed that all these species belonged to *G. maculatus*. Scott (1968) considered *alpinus* a separate species from *maculatus*. McDowall (1967, 1971b, 1972) reaffirms Stokell's position that *alpinus* is only a lacustrine population of *maculatus*. Campos (1974), studying the life cycle of *G. maculatus* differing in numbers of vertebrae, questioned the validity of assigning *alpinus* species status as later claimed by McDowall (1976a).

Recent studies have shed new light on the relationships within the Galaxiidae. Mitchell & Scott (1979) worked with 10 species of New Zealand galaxiids using 'muscle myogen protein' separated by starch gel electrophoresis. They found taxonomic groupings similar to those suggested by McDowall (1970).

Aplochitonidae

The family Aplochitonidae was formed by Günther (1864), who originally called it Haplochitonidae (McDowall 1969). Among other characteristics the aplochitonids have bodies without scales, an adipose dorsal fin and size dimorphism only for the genera *Aplochiton* and *Lovettia* (excluding *Prototroctes*). The genera *Aplochiton* and *Lovettia* show affinities to the Galaxiidae. Nelson (1972) includes them in the subfamily Galaxiinae together with *Galaxias*. Rosen (1974) places *Galaxias*, *Aplochiton* and *Lovettia* in the family Galaxiidae in the superfamily Salmonoidei. The majority of authors separate from this group the genus *Prototroctes*, which is elevated to the rank of family by McDowall (1969). The genus *Aplochiton* was described by Jenyns (1842), who described *Aplochiton zebra* of the Falkland Islands and *A. taeniatus* of Tierra del Fuego. Eigenmann (1928) described *Aplochiton marinus*, a species that Eigenmann also gives as pertaining to *taeniatus*, and which McDowall (1971a) considers synonymous to *taeniatus*. Although no sexual dimorphism has been recorded in the genus *Aplochiton*, we have found in *A. taeniatus* marked differences in the form of the urogenital papilla between mature adult male and female (Campos 1969). The genus *Lovettia* was established by McCulloch (1915) with only one species, *Lovettia seali* (Johnston) of Tasmania. This genus is characterized by its extreme sexual dimorphism in which mature adult males have their anal and urogenital aperture above the isthmus.

Prototroctidae

The family Prototroctidae was created by Hubbs (1953), and revalidated by McDowall (1969) who performed an osteological comparison of the genus *Prototroctes* with the genera *Lovettia*, *Retropinna*, *Aplochiton* and *Galaxias*. The family Prototroctidae is characterized by having a scaled body, no lateral line, adipose fin, and gut with 2 intestinal loops. Gosline (1960) suggested that the differences among *Aplochiton*, *Lovettia* and *Prototroctes* required separate families. McDowall (1969) concluded that the characteristics were sufficient for family rank. Nelson (1972) placed *Prototroctes* in a subfamily Retropinninae of the family Galaxiidae. Rosen (1974) placed *Prototroctes*, *Retropinna* and *Stokellia* in the family Retropinnidae of the superfamily Osmeroidei. McDowall (1969) argued that although *Prototroctes* has a great affinity with Retropinnidae, there are enough different characters to separate them. The genus *Prototroctes* was created by Günther (1864) and consists of two species, *Prototroctes oxyrhynchus* Günther of New Zealand and *Prototroctes marena* Günther of southeast Australia and Tasmania.

Retropinnidae

The family Retropinnidae was proposed by Regan (1913). Its characteristics include scales, a single left ovary, sexual dimorphism (nuptial tuberculae in the adult male), and an absence of lateral line. The Retropinnidae have two genera, *Retropinna* and *Stokellia*, and four species. The genus *Retropinna* Gill (1862) differs from *Stokellia* because of its short alveolar process which does not exclude the maxilla from the border of the mouth; sometimes the maxilla is dentated. The genus consists of three species, one from New Zealand, *Retropinna retropinna* (Richardson), one from Australia, *Retropinna semoni* (Weber), and a third from Tasmania, *Retropinna tasmanica* McCulloch. McDowall (1979) has recently revised the genus. The genus *Stokellia* Whitley (1955) differs from *Retropinna* because of its long premaxillar alveolar process which excludes the maxilla from the mouth. It has a single species, *Stokellia anisodon* (Stokell), restricted to New Zealand.

Ecology

The species of the southern salmoniform families are principally fresh water or anadromous. According to Rosen (1974) the Salmoniformes are fresh water fishes in which some groups (Argentinoidei and *Osmerus espelarnus*) have secondarily become marine. The relationship of the galaxioids to their environment is important in predicting whether they are capable of transcending the marine barriers that separate them in their distribution. Here I present what little information is available about their dependency on salt water.

In the Aplochitonidae the species lay their eggs in fresh water. We have observed the reproduction of *Aplochiton taeniatus* (Campos 1969). The mature adults approach land during winter (August) in great shoals, entering the outlets of lakes. Females lay an average of 3000 eggs which they slowly deposit on the substrate (rocks, boughs, sticks, etc.). The eggs remain stuck to the substrate and are covered with spermatic fluid through the quick movements of the male. The embryos hatch at temperatures of 10° C. The juveniles and young are found in the streams, protected from the water current among the stones, feeding mainly on aquatic insects. The adults seek lentic waters in the estuaries to feed. In estuaries of low salinity, such as the fiords of Southern Chile, the species seems abundant.

Aplochiton marinus Eigenmann is a species whose taxonomic position had been discussed since Eigenmann (1928) himself suggested the possibility that they may be the adults of *A. taeniatus*. This is not likely because adult *A. taeniatus* are found in lakes. The specimens we have collected from estuaries are larger than any known *A. taeniatus*, with a different coloring (with almost no stripes), and with a mouth shape which would correspond closer to *A. marinus*. Clearly, more studies are needed on these two species. *Aplochiton zebra* Jenyns has been found in small streams (Eigenmann 1928, McDowall 1971a), but precise data about their reproduction are not available. However, Eigenmann (1928) observed specimens laying eggs in March in a small stream that flows from Rinihue Lake. Smitt (1901) mentions that this species seems to spawn in March, but gives no further information.

The only other species related to the aplochitonids of South America is *Lovettia seali* of Tasmania, which is also a freshwater species. According to Blackburn (1950) the mature adults migrate from the sea to the rivers to spawn in autumn, depositing their eggs in much the same way as in *A. taeniatus*. Blackburn (1950) says that the larvae and juveniles migrate to the ocean. In summary, the aplochitonids are closely tied to the freshwater environment.

The family Galaxiidae has the greatest number of southern Salmoniformes, yet the biology of galaxiids is scarcely known. We have studied *Brachygalaxias bullocki*, a small fish restricted to fresh water (Fischer 1963, Campos 1972), in which reproduction takes place in winter (Campos 1972). The females have a low relative fecundity with an average of 100 eggs in specimens between 30 and 40 mm. Females attach eggs to aquatic plants and males fertilize them by releasing spermatic fluid nearby. The males seem to increase the brightness of their orange coloration at this time. The embryos hatch after 14 to 16 days in water temperatures between 9° and 12° C. Growth is direct, without a pigmented juvenile phase as in other galaxiids. We have not observed migratory behavior associated with reproduction in this species. The spawning areas are small slow streams or shallow lentic environments with aquatic vegetation. *G. pusillus* of Tasmania has a similar reproductive cycle to *B. bullocki*, according to Frankenberg (1974). This coincidence is interesting because *G. pusillus* has been thought to belong to *Brachygalaxias* (see preceding section).

Galaxias globiceps is known only from the holotypes and paratypes collected by Eigenmann (1928) from Los Alerces stream in Southern Chile. Nothing is known about the life history of this species. We assume that they are restricted to fresh water and that they live under the plants in slow streams.

Galaxias platei is the largest galaxiid in Chile reaching approximately 34 cm TL. This species has transparent juveniles (whitebait) smaller than the yellow *G. maculatus*. The pelagic juveniles are found in the shores of lakes and rivers. Juveniles kept in an aquarium begin to develop pigment and lose their pelagic behavior in favor of a demersal position when adults. These observations in the aquarium have been corroborated in nature where adults are found hidden under stones in the epirhithron of rivers of high-lands and at the bottom of lakes from the shore to deeper areas. We have collected specimens of 16 cm TL at 350 m depth in Rinihue Lake. The spawning areas of this species are not known. We have collected male and female specimens in advanced stages of maturation throughout all months of the year, coinciding with the permanent presence of juveniles along the shore of lakes. Moreover, we have found mature adults both in the fluvial habitat and in the deep waters of lakes. Females have a high relative fecundity with 1300 eggs at 7 cm, up to 30000 at 17 cm in specimens from the Panguipulli Lake. The egg diameter is smaller than other galaxiids of Chile. We have found the larval stages of *G. platei* in December (summer) in the Calle Calle (limnetic waters) and in Corral Harbour (brackish waters). In brackish waters we have also found juveniles which seem to migrate towards the river.

These observations lead us to conclude that the species is dependent on the sea, but restricted to estuaries. It is not known whether the eggs are laid both in brackish waters and fresh waters. The mere presence of larvae within a few hours of hatching in a brackish environment is not a clear proof because, since the species is demersal and lives under rocks in streams, it is possible that the larvae come from these streams. On the other hand, the number of vertebrae of the stocks in the estuary are not significantly different from the stocks in the river and lakes. The variation in vertebrae number seems to correlate with latitude (McDowall 1971b).

Galaxias maculatus is one of the most studied species of galaxiids in its range in Australia, New Zealand and South America. Knowledge of its biology has been an important argument in claiming the conspecificity of widely separated populations. In Chile, we have studied this species especially in the basin of the Valdivia River (Campos 1970, 1973, 1974). We have selected this basin because it is formed by 6 lakes of glacial origin, interconnected by short rivers, and the last lake, Rinihue, drains into the San Pedro River. The river has different names (Calle Calle, Valdivia) throughout its route of 200 km until it drains into the Pacific Ocean. The River Valdivia forms an estuary with a salinity of 0.5–30 ppt. The temperatures of the lakes and the river fluctuate between 8° and 21° C. The temperature of the mouth of the estuary (Corral Bay) is lower than that of the river because of the marine influence, especially during spring-summer.

G. maculatus has one population in the interior lakes, and another in the river. The populations possess different numbers of vertebrae. In the Calle Calle river we find both populations but in the period of reproduction during spring-summer, the mature adults segregate. The adults with the larger number of vertebrae migrate towards the estuary, and the others migrate up river. During the same period mature adults are found in the interior lakes. Their relative fecundity correlates with length, reaching about 400 eggs at 48 mm, and about 7400 eggs at 160 mm. We have not observed the eggs laid in nature as described from New Zealand (Benzie 1968) and some Australian lakes (Pollard 1971). However, in freshwater aquariums, the females lay for several days, slowly expelling the eggs and dropping them to the bottom from a mid-water position. The maximum number of eggs we observed a female lay was 1097 in February (summer). Males remained at the bottom of the aquarium expelling their seminal liquid near the accumulation of eggs. Embryonic development took 16 days at a temperature of 17° C and the newly hatched embryos were 5.6 mm long. The spawning behavior is somewhat different from that described from New Zealand where the eggs adhere to aquatic plants. In nature, the free embryos and larvae remain near their spawning grounds. We have collected larvae in the Harbour of Corral and up river. From the larvae, transparent diadromous juveniles (whitebait) develop, which remain in the harbour and even approach the coast in order to migrate into the river later on. The larger migrations from the sea into the river are carried out in winter. In freshwater these juveniles transform, their bodies becoming pigmented. The cycle is annual, but it seems that some adults live more than a year. Similar cycles have been described in New Zealand (McDowall 1968, Benzie 1968), Australia (Frankenberg 1969), and Tasmania (Pollard 1966).

One of the controversial problems of *G. maculatus* is the taxonomic status of the lacustrine populations. McDowall (1972) has compared lacustrine populations from New Zealand, Australia and South America. Several of these populations are considered landlocked or restricted to freshwater, and have lost their connection with the sea. These populations spawn in the streams associated with lakes as has been shown by Pollard (1971) in Modewarre. The lacustrine populations have significantly fewer vertebrae than the diadromous ones but, according to McDowall (1972), are not taxonomically different from them. Our data (Campos 1974) showed a bimodal curve separating the populations with a low number of vertebrae from those with a high number, and distinct reproductive habitat and migratory behavior. This suggested that populations with a low number of vertebrae corresponded to *Galaxias alpinus* (Jenyns), and those with a high number of vertebrae to *Galaxias maculatus;* reproductive isolation can be presumed. McDowall (1976a) writes extensively about the synonymy of *Galaxias alpinus* with *G. maculatus*, claiming that the lacustrine populations are not a separate species. It is not my intention to retain the species *G. alpinus* as separate; new research should indicate their taxonomic position.

Reproductive isolation is recognized or deduced from isolating mechanisms (Dobzhansky 1937), which are defined by Mayr (1965) as '... biological properties of individuals that prevent the interbreeding of populations that are actually or potentially sympatric'. Mayr's opinion is that the less frequently a male and female encounter one another, the less their probability of mating, so that the selection of habitat can constitute a very efficient isolating mechanism. Consequently, if the mature adults with high and low numbers of vertebrae have selected different habitats in the same river to reproduce, as our data reveal (Campos 1974), it does not seem a 'simplification' (McDowall's term) to suppose that a reproductive isolation mechanism is present, at least for the majority of the population. If no isolating mechanism were present, we should have found mature adults of high and low number of vertebrae in the estuary and up river or in the lakes. It is possible that both populations cross, as McDowall (1972) says, but my observations show that the two populations follow different migratory routes. I agree with McDowall (1972) that the number of vertebrae is a result of the habitat, but I do not agree that its value as an indirect indicator of taxonomic status should be rejected for that reason.

Variation in the number of vertebrae in fishes caused by environmental factors has been widely studied, but several authors acknowledge the effect of the genotype on this variation (Fowler 1970, Ali & Lindsey 1974). Lindsey (1975) also discusses the contributions of the number of vertebrae to the taxonomy of fishes. There is evidence from populations of *Oncorhynchus keta* and *Salmo gairdneri* that intraspecific variations in the number of vertebrae persist in offspring which are cultivated under identical conditions (Kubo 1956, Mottley 1937). So far there are no published experiments that reveal what are the abiotic factors that maintain the variations between the lacustrine and diadromous populations of *G. maculatus*, and whether this characteristic would be kept if both populations were cultivated in the same conditions. All lacustrine populations are morphologically very similar in the different continents (Campos 1979). The populations were supposed to have a diadromous origin as postulated by McDowall (1972). The interior lakes of Chile were formed during the last glaciation, and possibly the diadromous populations were in contact with the streams of the coastal cordillera which were not affected by the glaciers before those glaciations. It must be kept in mind that many salmoniform fishes home to their spawning places. This can keep both populations separated at their spawning places and initiate a process of speciation. The protein variability in lacustrine populations is higher than in the diadromous populations (unpublished data).

Zoogeography

The galaxioids are distributed in South Africa

(Cape Peninsula), South America and the Falkland Islands, New Zealand, Australia, Tasmania and New Caledonia (see Fig. 1). Of the four families, Galaxiidae is distributed throughout the area, the Retropinnidae and Prototroctidae in New Zealand, Australia and Tasmania, and the Aplochitonidae in South America and Tasmania. The greatest diversity of galaxioids (all 4 families, 6 genera and 18 species) occurs in Tasmania. In continental Australia only three families are found, 4 genera and 11 species, 5 species being shared with Tasmania. In New Zealand three families, 5 genera, and 16 species are found. In South America there are only 2 families, with 3 genera and 6 species. In South Africa only the family Galaxiidae is present, with only 1 species of the genus *Galaxias*. In New Caledonia, as in South Africa, the family has but an endemic genus, *Nesogalaxias*, with 1 species.

The southern families originated with the osmeroid and salmonoid groups of the northern hemisphere (Fig. 1). Rosen (1974) concludes that the salmoniforms could have been in Pangaea when the ancestral continent broke up into Laurasia and Gondwanaland 180 million years ago. According to this historical explanation, the ancestral stocks would be as old as the date at which they separated. The galaxiids and aplochitonids could have their origin in a stock of salmonoids, and the retropinnids and prototroctids in an osmeroid stock. In the only viable alternative, Hubbs (1953) proposed that the Galaxiidae were formed in the upper Tertiary. In this case dispersal from north to south, crossing the tropics must be assumed. Such a dispersal is unlikely, given what is known about climates in the tropics during the Tertiary. It would seem that the historical rather than the dispersal explanation provides greater insight into the origin of the families in different ancestral groups of the northern hemisphere.

Paleontological evidence of the galaxioids is very scarce and restricted to New Zealand. Oliver (1936) found a petrified *Galaxias* in Fraser's ravine in the Kaikorai region near the city of Dunedin. Stokell (1945) compared this fossil with the extant species and concluded that its characteristics were a combination of many species such as *Galaxias attenuatus*, *G. alepidotus*, *G. fasciatus*, *G. paucispondylus* and *G. prognathus*. The associated organisms show the fossil had lived in freshwater of the Pliocene of New Zealand. Later, Travis (1965) found 2 fossils of *Galaxias* in Fouldern Hills, near Middlemarch, Otago. McDowall (1976b) has studied these fossils and classifies them in the extant species *G. brevipinnis* (for the Dunedin specimen) and *G. vulgaris* (for the Fouldern Hill specimen). McDowall (1970) discounts the value of these fossils in determining the age of the family. He also postulates that this fish fauna may have reached New Zealand in the Upper or Middle Tertiary.

The greatest controversy lies in explaining how the actual zoogeographic patterns and the phylogenetic relationship of galaxioids were formed. The theories can be separated into ecological (i.e. via dispersion) versus historical (i.e. vicariant events). The ecological theories present oceanic dispersion as the principal mechanism. Some proponents of marine dispersion also suppose that the galaxioids had marine ancestors, that they travel long distances to reach new habitats, and that some species can reproduce in the sea (Myers 1949, Darlington 1957, 1965).

The majority of these authors lacked reliable knowledge about galaxioid distributions. This is not true of McDowall (1964, 1966, 1970, 1973b) who has contributed enormously to our knowledge about galaxioids and who is at present the main proponent of oceanic dispersion of the galaxiids to explain their present distributional patterns. McDowall (1970) thinks that the center of origin of the galaxiids was in Australasia. He argues that the distribution of the galaxioids is the result of their past and present capacity to disperse in the sea, crossing the oceanic barriers that separate their distribution. McDowall's principal argument has been, therefore, to show their dependency on the sea through the presence of a juvenile phase (whitebait) in several species of diadromous *Galaxias*. He emphasizes that through marine stages dispersion was transoceanic, Australia becoming the center of dispersion, going through successive invasions to New Zealand through the 'east Australian Ocean Current'. McDowall (1970) maintains that through the 'west wind drift' or the occidental antarctic current the galaxioids would have reached South

America and finally South Africa. As supporting evidence he emphasizes the decrease in the number of species from Australia to New Zealand, South America and South Africa. McDowall (1975) claims to have found juveniles of *Galaxias* (*Galaxias maculatus* and possibly *G. fasciatus* and *G. brevipinnis*) up to approximately 704 km off the coast of New Zealand. Earlier he (McDowall 1966) proposed that 'the range of one of these species – *G. maculatus* – suggests that this dispersal continues', which would imply that individuals of *maculatus* are presently reaching South American coasts.

I do not share the belief in transoceanic dispersion of galaxioids and have questioned it in the specific case of *Galaxias maculatus* (Campos 1973). This species, like the majority of the northern hemisphere salmonoids, has a freshwater phase and an estuarine or marine phase. The two phases are united by anadromous migratory reproductive behavior. If one phase is omitted it is the marine one, similar to most species of Salmonids. These phases are united by a strong homing instinct. I do not think that there is a marine phase in a strict sense, but rather an estuarine phase that reproduces in saltwater, with the juveniles feeding near the coast before entering the rivers (Campos 1973). Stokell (1972) feels that the term 'catadromous' does not apply well to these species and that their migration is more of a movement within the freshwaters.

McDowall (1968) says that in *G. maculatus* the spawning grounds can be in fresh or salt water; the important aspect is the presence of tidal influence. Benzie (1968) revealed spawning in estuarine areas, hence not typical marine areas, such as the intertidal coast. Rosen (1974) also doubts that spawning grounds of *G. maculatus* are strictly marine. Pollard (1971) studied the spawning habits of landlocked *G. maculatus* in Modewarre lake in Australia. The brackish waters of this lake have salinities ranging from 3 to $5^{0}/_{00}$.

Another problem with the dispersal hypothesis is the great similarity among *Brachygalaxias bullocki* of Chile, *Galaxias zebratus* of South Africa and the *B. pusillus* and *B. nigrostriatus* of Tasmania (Scott 1942, Stokell 1954). These species, because of their biology, seem to be restricted to freshwater, at least in the case of *B. bullocki*. McDowall (1973a) says, 'Inclusion of *G. nigrostriatus* and *G. pusillus* in *Brachygalaxias* would weigh the zoogeographic and phylogenetic relationship of these two species rather more strongly in the direction of common ancestry and close relationship to *B. bullocki* than existing evidence seems to warrant'.

In Aplochitonidae, knowledge about the reproduction of the species *Aplochiton* does not support McDowall's theory (1971a) of a transoceanic dispersion based upon a marine phase. *Aplochiton zebra* and *A. taeniatus* reproduce in freshwater, and the presence of adults in the estuaries cannot be considered evidence of a marine phase. *Lovettia* reproduces in freshwater and although juveniles and adults are found near the coastline, the genus is restricted to Tasmania. Rosen (1974, 1978) concludes that the species are primarily freshwater, and those considered secondarily freshwater are continental groups tolerant to salt water, that can only cross short marine distances. This would not explain the wide distribution from Australia to South Africa.

Historical biogeography is identified by Rosen (1978) as 'a search for historical connection between biotas in time and space'. Rosen (1974) established a similarity of the distributional pattern of the galaxioids to the Gondwana pattern of the midges studied by Brundin (1966). Brundin carried out a complete monographic study of the midges adapted to the cold in the southern hemisphere, with an elaborated phylogeny of the groups, and a comparison of a biological area cladogram with a geological area cladogram of the region. Rosen considered the geological area cladogram of disruption of the Gondwana continent comparable to the biological area cladogram of the galaxioids, even though there is not a precise phylogeny of this group. 'A geological area-cladogram that corresponds in part or whole with a biological pattern is simply adopted as an explanation of the pattern in the sense of providing a best current estimate of the historical factors that induced the biological pattern' (Rosen 1978, p. 186).

My studies (Campos 1969, 1970, 1972, 1973, 1974, 1979) have led me to conclude that the best explanation for the disjoint distribution of the galaxioid fishes (Galaxiidae, Aplochitonidae, Retropinnidae and Prototroctidae) is the disruption of the hypothetical continent of Gondwana. Generally speaking, what is needed is a detailed phylogeny of the southern salmoniforms to achieve a better explanation of their distributional patterns. Within a general Gondwana distribution, the galaxioids are merely a part of a pattern, and this makes the search for the historical rather than the ecological explanation more interesting.

References cited

Ali, M.Y. & C.C. Lindsey. 1974. Heritable and temperature-induced meristic variation in the medaka *Oryzias latipes*. Can. J. Zool. 52: 959–976.

Barnard, K.H. 1943. Revision of the indigenous freshwater fishes of the S.W. Cape region. Ann. S. Afr. Mus. 36: 101–262.

Benzie, V. 1968. Some ecological aspects of the spawning behavior and early development of the common whitebait *Galaxias maculatus attenuatus* (Jenyns). N.Z. Ecol. Soc. 15: 31–39.

Blackburn, M. 1950. The Tasmania whitebait, *Lovettia seali* (Johnston), and the whitebait fishery. Austr. J. Mar. Freshwat. Res. 2: 155–198.

Brundin, L. 1966. Transantarctic relationships and their significance, as evidenced by chironomid midges. K. Svenska Vetenskaps-akad. Handl. 4: 1–472.

Campos, H. 1969. Reproduction of *Aplochiton taeniatus* Jenyns. Bol. Mus. Hist. Nat. (Chile) 29: 207–222. (In Spanish).

Campos, H. 1970. *Galaxias maculatus* (Jenyns) in Chile, with special reference to its reproduction. Bol. Mus. Nac. His. Nat. (Chile) 31: 5–20 (In Spanish).

Campos, H. 1972. Kariology of three galaxiid fishes *Galaxias maculatus*, *G. platei* and *Brachygalaxias bullocki*. Copeia 1972: 368–370.

Campos, H. 1973. Migration of *Galaxias maculatus* (Jenyns) (Galaxiidae, Pisces) in Valdivia Estuary, Chile. Hydrobiologia 43: 301–312.

Campos, H. 1974. Population studies of *Galaxias maculatus* (Jenyns) (Osteichthyes: Galaxiidae) in Chile with reference to the number of vertebrae. Stud. Neotrop. Fauna 9: 55–76.

Campos, H. 1979. Multivariate analysis of the taxonomy of the fish family Galaxiidae. Zool. Anz. 202: 280–288.

Chapman, W.M. 1944. The osteology and relationships of the South American fish, *Aplochiton zebra* Jenyns. J. Morphol. 75: 149–165.

Cuvier, G.L.C.F.D. & A. Valenciennes. 1846. Des Galaxies. pp. 340–357. *In*: Histoire Naturelle des Poissons, 18, Levrault, Paris.

Darlington, P.J., Jr. 1957. Zoogeography, the geographical distribution of animals. John Wiley and Sons, New York. 675 pp.

Darlington, P.J., Jr. 1965. Biogeography of the southern end of the world. Harvard Univ. Press, Cambridge. 236 pp.

Dobzhansky, T. 1937. Genetics and the origin of species. 1st Edition. Columbia University Press, New York. 364 pp.

Eigenmann, C.H. 1928. The freshwater fishes of Chile. Mem. Nat. Acad. Sci. 22: 1–80.

Fischer, W. 1963. Die Fische des Brackwasser Gebietes Lenga bei Concepcion, Chile. Int. Revue Ges. Hydrobiol. 48: 491–511.

Fowler, J.A. 1970. Control of vertebral number in teleosts: an embryological problem. Q. Rev. Biol. 45: 148–167.

Frankenberg, R.S. 1969. Studies on the evolution of galaxiid fishes with particular reference to the Australian fauna. Ph. D. Thesis, University of Melbourne, Melbourne.

Frankenberg, R. 1974. Native freshwater fish. *In*: W.D. Williams (ed.) Biogeography and Ecology in Tasmania, Dr W. Junk, The Hague.

Gill, T.N. 1862. On the subfamily of Argentininae. Proc. Acad. Nat. Sci. Phila. 14: 14–15.

Gill, T. 1893. A comparison of antipodal faunas. Mem. Nat. Acad. Sci. 6: 91–124.

Gosline, W.A. 1960. Contribution toward a classification of modern isospondylous fishes. Bull. Brit. Mus. (Nat. Hist.) Zool. 6: 321–365.

Greenwood, P.H., E.D. Rosen, S.H. Weitzman & G.S. Myers. 1966. Phyletic studies of teleostean fishes with a provisional classification of living forms. Bull. Amer. Mus. Nat. Hist. 131: 339–456.

Günther, A. 1864. Catalogue of the fishes in the collection of the British Museum. Vol. 5. British Museum, London. pp. 382–383.

Günther, A. 1866. Catalogue of fishes in the British Museum. Vol. 6. British Museum, London. 368 pp.

Günther, A. 1867. On a new form of mudfish from New Zealand. Ann. Mag. Nat. Hist. Ser. 3: 305–309.

Günther, A. 1880. An introduction to the study of fishes. Black, Cambridge.

Hubbs, C.L. 1953. Antitropical distribution of fishes and other organisms. Proc. 7th Pac. Sci. Congr. 3: 324–329.

Jenyns, L. 1842. The zoology of H.M.S. Beagle during 1832–1836. Pt. 4. Fish. Smith Elde, London, 172 pp.

Kubo, T. 1956. Peculiarity of population of chum salmon in the Shiriuchi River in respect to the vertebral count. Bull. Fac. Fish. Hokkaido Univ. 6: 266–270.

Lindsey, C.C. 1975. Pleomerism, the widespread tendency among related fish species for vertebral number to be correlated with maximum body length. J. Fish. Res. Board Can. 32: 2453–2469.

Mayr, E. 1965. Animal species and evolution. Belknap Press, Cambridge. 797 pp.

McCulloch, A.R. 1915. Notes on and description of Australian fishes. Proc. Linn. Soc. New South Wales 40: 259–277.

McDowall, R.M. 1964. The affinities and derivation of the New Zealand freshwater fish fauna. Tuatara 12: 56–67.

McDowall, R.M. 1966. Further observations on *Galaxias* whitebait and their relation to the distribution of the Galaxiidae.

Tuatara 14: 12–18.
McDowall, R.M. 1967. Some points of confusion in galaxiid nomenclature. Copeia 1967: 841–843.
McDowall, R.M. 1968. *Galaxias maculatus* (Jenyns) the New Zealand whitebait. Fish. Res. Bull. N.Z. Marine Dept. Wellington 2: 1–84.
McDowall, R.M. 1969. Relationships of galaxiid fishes with a further discussion of salmoniform classification. Copeia 1969: 796–824.
McDowall, R.M. 1970. The galaxiid fishes of New Zealand. Bull. Mus. Comp. Zool. 139: 341–431.
McDowall, R.M. 1971a. Fishes of the family Aplochitonidae. J. Royal Soc. N.Z. 1: 31–52.
McDowall, R.M. 1971b. The galaxiid fishes of South America. Zool. J. Linn. Soc. 50: 33–73.
McDowall, R.M. 1972. The species problem in freshwater fishes and the taxonomy of diadromous and freshwater populations of *Galaxias maculatus* (Jenyns). J. Royal Soc. N.Z. 2: 325–367.
McDowall, R.M. 1973a. Limitation of the genus *Brachygalaxias* Eigenmann, 1928 (Pisces: Galaxiidae). J. Royal Soc. N.Z. 3: 193–197.
McDowall, R.M. 1973b. The status of the South African galaxiids (Pisces: Galaxiidae). Annals of the Cape Province Museum 9: 91–101.
McDowall, R.M. 1975. Occurrence of galaxiid larvae and juveniles in the sea. N.Z. J. Marine & Freshwater Res. 9: 1–9.
McDowall, R.M. 1976a. The taxonomic status of the *Galaxias* populations in the Rio Calle-Calle, Chile (Pisces: Galaxiidae). Stud. Neotropical Fauna 11: 173–177.
McDowall, R.M. 1976b. Notes on some *Galaxias* fossils from the Pliocene of New Zealand. J. Royal Soc. N.Z. 6: 17–22.
McDowall, R.M. 1978. Generalized tracks and dispersal in biogeography. Syst. Zool. 27: 88–104.
McDowall, R.M. 1979. Fishes of the family Retropinnidae (Pisces: Salmoniformes) – A taxonomic revision and synopsis. J. Royal Soc. N.Z. 9: 85–121.
McDowall, R.M. & R.S. Frankenberg. 1981. The galaxiid Fishes of Australia (Pisces: Galaxiidae). Rec. Aust. Mus. 33: 443–605.
Merriles, M.J. 1975. Karyotype of *Galaxias maculatus* from New Zealand. Copeia 1975: 176–178.
Mitchell, C.P. & D. Scott. 1979. Muscle myogens in the New Zealand Galaxiidae. N.Z. J. Marine & Freshwater Res. 13: 285–294.
Mottley, C.Mc.C. 1937. The number of vertebrae in trout (*Salmo*). J. Biol. Board Can. 3: 169–176.
Müller, J. 1844. Über den Bau und die Grenzen der Ganoiden und über das natürliche System der Fische. Abhandl. Akad. Wiss. Berl. 1844: 117–216.
Myers, G.S. 1949. Salt tolerance of freshwater fish groups in relation to zoogeographical problems. Bijd. tot Dierk. 28: 315–322.
Nelson, G.J. 1972. Cephalic sensory canals, pitlines, and the classification of esocoid fishes, with notes on galaxiids and other teleosts. Amer. Mus. Novitates 2492: 1–49.
Ogilby, J.D. 1899. Contributions to Australian ichthyology. Proc. Linn. Soc. New South Wales 24: 154–186.
Oliver, W.R.B. 1936. The Tertiary flora of the Kaikorai Valley, Otago, New Zealand. Trans. Royal Soc. N.Z. 66: 284–304.
Pollard, D.A. 1966. Land-locking in diadromous salmonoid fishes with special reference to the common jolly-tail (*Galaxias attenuatus*). Australian Soc. Limn. Newslett. 5: 13–16.
Pollard, D.A. 1971. The biology of a landlocked form of the normally catadromous salmoniform fish *Galaxias maculatus* (Jenyns). Aust. J. Mar. Freshwat. Res. 22: 91–123.
Regan, C.T. 1905. A revision of the fishes of the family Galaxiidae. Proc. Zool. Soc. London. 2: 363–384.
Regan, C.T. 1913. Antarctic fishes of the Scottish National Antarctic Expedition. Trans. Royal Soc. Edinburgh 49: 229–292.
Rosen, D.E. 1974. Phylogeny and zoogeography of salmoniform fishes and relationship of *Lepidogalaxias salamandroides*. Bull. Amer. Mus. Nat. Hist. 153: 265–326.
Rosen, D.E. 1978. Vicariant patterns and historical explanation in biogeography. Systematic Zoology 27: 159–188.
Scott, E.O.G. 1935. On a new genus of fishes of the family Galaxiidae. Pap. Proc. Royal Soc. Tasmania 1934: 41–46.
Scott, E.O.G. 1936. Observation on fishes of the family Galaxiidae. Part I. Pap. Proc. Royal Soc. Tasmania 1935: 85–112.
Scott, E.O.G. 1942. Description of Tasmanian mud trout, *Galaxias* (*Galaxias*) *upcheri* sp. nov.: with a note on the genus *Brachygalaxias* Eigenmann 1928. Records of the Queen Victoria Museum, Launceston, Set 1: 51–57.
Scott, E.O.G. 1966. The genera of Galaxiidae. Austral. Zool. 13: 244–258.
Scott, E.O.G. 1968. Certain nomenclatural proposals in Galaxiidae: a rejoinder. Records of the Queen Victoria Museum Lauceston, Set 29: 1–10.
Scott, E.O.G. 1971. On the occurrence in Tasmania and on Flinders Island of *Brachygalaxias* Eigenmann 1928 (Pisces, Galaxiidae) with a description of two new subspecies. Records of the Queen Victoria Museum, Launceston, Set 37: 1–14.
Smitt, F.A. 1901. Poissons d'eau douce de la Patagone recueillis par E. Nordenskjöld 1898–99. Bih. K. Svenska Vetensk-Akad. Handl. 26 4: 1–31.
Stokell, G. 1945. The systematic arrangement of the New Zealand Galaxiidae. I. Generic and subgeneric classification. Trans. Royal Soc. N.Z. 75: 124–137.
Stokell, G. 1949. The systematic arrangement of the New Zealand Galaxiidae. II. Specific classification. Trans. Royal Soc. N.Z. 77: 427–496.
Stokell, G. 1950. A revision of the genus *Paragalaxias*. Records of the Queen Victoria Museum, Launceston, Set 3: 1–4.
Stokell, G. 1954. Contributions to galaxiid taxonomy. Trans. Royal Soc. N.Z. 82: 411–418.
Stokell, G. 1966. A preliminary investigation of the systematics of some Tasmanian Galaxiidae. Pap. Royal Soc. Tasmania 100: 73–79.
Stokell, G. 1972. Freshwater and diadromous fishes of New Zealand. Canterbury Mus. Bull. 5. 48 pp.
Travis, C. 1965. The geology of the Slip Hill area east of Middlemarch, Otago. M.S. Thesis, University of Otago,

Otago.
Whitley, G.P. 1935. Whitebait. Vict. Nat. Melbourne 52: 41–51.
Whitley, G.P. 1955. Sidelights on New Zealand ichthyology. Aust. Zool. 12: 110–119.
Whitley, G.P. 1956. The story of Galaxias. The Australian Mus. Mag., March 15, 1956: 30–34.
Weitzman, S.H. 1967. The origin of the stomiatoid fishes with comments on the classification of salmoniform fishes. Copeia 1967: 507–540.

Corydoras marmoratus from Steindachner (1879), Über einige neue und seltene Fisch-Arten aus den K.K. Zoologischen Museen zu Wien, Stuttgart und Warschau. Denkschr. Akad. Wiss. Wien 41: 1–52.

On measuring niches and not measuring them

Thomas M. Zaret[1] & Eric P. Smith[2]
[1] *Institute for Environmental Studies and Department of Zoology, University of Washington, FM-12, Seattle, WA 98195, U.S.A.*
[2] *Department of Statistics, Virginia Polytechnic Institute and State University, Blacksburg, VA 240661, U.S.A.*

Keywords: Niche overlap, Statistics, Fishes, Evolution, Competition, Ecology, Speciation, Historical ecology

What birds can have their bills more peculiarly formed than the ibis, the spoonbill, and the heron? Yet they may be seen side by side, picking up the same food from the shallow water on the beach; and upon opening their stomachs, we find the same little crustacea and shellfish in them all.

From A.R. Wallace, 'A Narrative of Travels on the Amazon and Rio Negro', second edition of 1899, Dover Publications. p. 59.

Synopsis

Niche overlap measures have been used extensively by ecologists, but until recently it was not possible to evaluate whether calculated overlap values differed significantly from chance expectations. We provide test statistics and a method for calculating confidence intervals for 3 overlap measures based on: (1) the likelihood ratio statistic; (2) the chi-square goodness-of-fit statistic; and (3) the Freeman-Tukey statistic. We also provide a means of calculating confidence intervals for 3 other commonly used measures, (4) the percentage similarity index; (5) Morisita's adjusted measure; and (6) an information theory index.

Comparing the measures for different values of N (total sample size of all resources) and r (number of different resource categories), we recommend an N of at least 100 for accurate results. Above this value, changes in N and r have a lessened effect on the results. Of the 6 measures, we recommend the overlap equation based on the Freeman-Tukey statistic, primarily because the variance estimates do not depend on the resource data set, but on the measure itself. Using our methods one can (a) compare a calculated overlap value with any specified value (e.g., does it differ from 1.00, 0.9, etc.), (b) compare overlap values for the same species at different times (e.g., day versus night, day one versus day two, etc.).

The second part of this contribution considers the origin of morphological diversity. A simple model suggests that interspecific competition is not necessary for the origin of specific differences among closely related taxa. Species diversity can occur simply from: (1) separation of one population into genetically isolated units; (2) sufficient time for the evolution of species-specific differences; and (3) sympatry of the original population. We discuss the need to distinguish between long-term, population isolation leading to speciation, versus short-term adaptation of closely related taxa. Only by evaluating the role of time can we interpret the meaning of morphological diversity.

Thomas M. Zaret (ed.), Evolutionary ecology of neotropical freshwater fishes. ISBN 90 6193 823 6

Introduction

The concept of 'the niche' is one of the cornerstones of modern ecological theory (see Whittaker & Levin 1975 for most complete recent treatment). Yet this concept still meets with skepticism among field scientists studying tropical freshwater fish communities. For example, Fryer & Iles (1972, esp. pp. 276–296), discuss numerous African species living in syntopy, whose diets appear indistinguishable. Others have similar information for neotropical communities (Goulding 1980). Tropical fish communities repeatedly raise nettlesome questions about the concept of niche, and whether, in spite of obvious differences among the species, many tropical fishes are not simply opportunistic generalists.

This paper contains two parts, loosely related by their focus on the ecological concept of niche. Part I addresses the question of how best to evaluate niche overlap measures. Even though there has been a great deal of research on measuring niche overlap, it is only in the last few years that there have been serious attempts to evaluate calculated measures of niche overlap using a statistical basis. In the second part of this paper we address a related question. When we find two species with little overlap in their 'niche', be it aspects of diet, behavior, habitat use, etc., is this the result of ecological interactions between the two species (e.g., character displacement) or differences evolved independently, and the species only secondarily sympatric? For tropical freshwater fishes this is analogous to asking whether they have well defined niches or are opportunistic and their diets merely reflect what they have, within limits, been able to acquire. A brief model is described which shows that we need not rely on theories of niche partitioning to explain spatial, temporal, or dietary overlaps among closely related species.

Part 1: Niche overlap

Measures of overlap

The development and use of niche overlap measures has been a common focus of study among ecologists interested in competition theory (e.g., MacArthur 1968, Cody 1974, Schoener 1974). The interpretation of these measures, however, particularly the relationship between calculated overlap values and presumed competitive pressures, is still receiving considerable attention (Hurlbert 1978, Abrams 1980). There are two problems associated with the interpretation of niche overlap measures. First, there may be no interpretable relation between overlap measurement and competition intensity if, for example, observations of low niche overlap are due to a response to past competition (see discussion and references in Zaret & Rand 1971, Lawlor 1980). This is a conceptual problem which cannot be resolved easily. A second problem concerns the interpretation of specific values calculated from niche overlap measures. It is only recently that these values have been critically evaluated using statistical tests or other criteria to indicate whether a calculated value differs from what is to be expected. Petraitis (1979) presents a measure of niche overlap which has good statistical properties and is interpretable as the likelihood that the usage distributions of two species are the same. Ricklefs & Lau (1980), Linton et al. (1981) and Smith & Zaret (1982) present results on bias in estimating measures of overlap. Ricklefs & Lau (1980) and Linton et al. (1981) also present results for variance and estimation, using Monte Carlo methods.

In this part we present methods for estimating the variance of six measures of niche overlap, assuming a multinomial model of resource use. The estimates of variance may be used to compute confidence intervals for the measures, and to test whether the measures are equal to a fixed value. We consider four commonly used measures:

(1) The proportional similarity measure (Renkonen 1938)

$$PS = \sum_{j=1}^{r} \min(p_{1j}, p_{2j}) = 1 - \tfrac{1}{2}\sum_{j=1}^{r} |p_{1j} - p_{2j}|, \quad (1)$$

where p_{1j} is the proportion of resource j used by species 1, p_{2j} is the proportion used by species 2, and r is the number of resources used by the species.

(2) Morisita's (1959) measure as adjusted by Horn (1966)

$$C_1 = 2 \sum_{j=1}^{r} p_{1j}p_{2j} \Big/ \left(\sum_{j=1}^{r} p_{1j}^2 + \sum_{j=1}^{r} p_{2j}^2 \right) \quad (2)$$

(3) Horn's (1966) information measure

$$R_0 = \left[\sum_{j=1}^{r} \{(p_{1j} + p_{2j})\, ln(p_{1j} + p_{2j}) - - p_{1j}\, ln p_{1j} - p_{2j}\, ln p_{2j}\} \right] / 2 ln(2). \quad (3)$$

(4) Petraitis' (1979) likelihood measure

$$L = \exp \left\{ \frac{1}{N} \sum_{i=1}^{2} \sum_{j=1}^{r} N_i p_{ij} (ln c_j - ln p_{ij}) \right\}, \quad (4)$$

where c_j = use of resource j assuming $p_{1j} = p_{2j}$, N_i = total resource use by species i, $N = N_1 + N_2$.

Additionally, two measures which have nice statistical properties are considered.

(5) The measure based on the chi-square goodness-of-fit statistic (van Belle & Ahmad 1974)

$$CX = 2 \sum_{j=1}^{r} p_{1j}p_{2j}/(p_{1j} + p_{2j}). \quad (5)$$

(6) The measure of Matusita (1955)

$$FT = \sum_{j=1}^{r} (p_{1j}p_{2j})^{\frac{1}{2}}. \quad (6)$$

The measures CX and FT are not commonly used in ecological studies. The CX measure may be motivated as the two sample extension of the niche breadth measure of Hurlbert (1978). Some justification for the use of the FT measure is given in Smith (1982). Petraitis' measure has a range of (0.5–1.0) while the other five measures are between zero and one. This work differs from Ricklefs & Lau (1980) and Linton et al. (1981) in that analytical methods are used to estimate the variance rather than simulation methods.

Statistical assumptions

Our statistical assumptions follow Petraitis (1979), Ricklefs & Lau (1980) and Smith & Zaret (1982). We assume that the usage vectors $\underset{\sim}{n}_i = (n_{i1}, n_{i2}, \ldots, n_{ir})$ are multinomial $M(N_i, \underset{\sim}{p}_i)$. The multinomial model may be used for both discrete and continuous data. If the data are discrete, for example, prey items in the diet, the amount of prey type j found in predator species i may be modelled as multinomial. If the data are continuous, for example, prey length, then one may render the data discrete by dividing the continuous variable length into length classes. Then n_{ij} is the number of prey of length class j taken by species i. In estimating overlap, one uses the maximum likelihood estimate of p_{ij} i.e. $\hat{p}_{ij} = n_{ij}/N_i$. The common usage c_j is estimated by $\hat{c}_j = (n_{1j} + n_{2j})/(N_1 + N_2)$.

Variance estimation and statistical tests

Let y denote a measure of overlap. We first consider results for tests of the form $H_0: y = y_0 \neq 1$, i.e. testing overlap equal to a given value. In this case, the statistical approach is to use the fact that as N_i becomes large, p_{ij} is approximately normal with variance $p_{ij}(1 - p_{ij})/N_i$. Since y is a function of p_{ij}, the variance of y can be estimated using the delta method. Additionally, y is approximately normally distributed (Seber 1973). Although the formula presented in Goodman & Kruskal (1979, p. 139) could be used as a formula for the variance of y, an extension of a formula in Fisher (1958, p. 309) (corrected by Samuel-Cahn 1975) has a simpler form;

$$\hat{V}(\hat{y}) = \sum_{i=1}^{2} 1/N_i \left\{ \sum_{j=1}^{r} \hat{p}_{ij} \left(\frac{\partial \hat{y}}{\partial p_{ij}} \right)^2 - - \left(\sum_{j=1}^{r} \hat{p}_{ij} \frac{\partial \hat{y}}{\partial p_{ij}} \right)^2 \right\}. \quad (7)$$

For the six measures considered, we have

$$\hat{V}(P\hat{S}) = \tfrac{1}{4} \left[\sum_{i=1}^{2} 1/N_i \left\{ \sum_{j=1}^{r} \hat{p}_{ij}\, I_j^2 - - (\sum \hat{p}_{ij}\, I_j)^2 \right\} \right], \quad (8)$$

$$\text{where } I_j = \begin{cases} 1 & \text{if } \hat{p}_{1j} < \hat{p}_{2j} \\ 0 & \hat{p}_{1j} = \hat{p}_{2j} \\ -1 & \hat{p}_{1j} > \hat{p}_{2j} \end{cases}$$

$$\hat{V}(\hat{C}_1) = \sum_{i=1}^{2} 1/N_i \left\{ \sum_{j=1}^{r} \hat{p}_{ij} b_{ij}^2 - - \left(\sum_{j=1}^{r} \hat{p}_{ij} b_{ij} \right)^2 \right\} \quad (9)$$

where $b_{1j} = 2(\hat{p}_{2j} - \hat{C}_1 \hat{p}_{1j}) / (\sum \hat{p}_{1j}^2 + \sum \hat{p}_{2j}^2)$

and $b_{2j} = 2(\hat{p}_{1j} - \hat{C}_1\hat{p}_{2j})/(\sum \hat{p}_{1j}^2 + \sum \hat{p}_{2j}^2)$.

$$\hat{V}(\hat{R}_0) = \sum_{i=1}^{2} 1/N_i \left[\sum_{j=1}^{r} \hat{p}_{ij} \{ln(\hat{p}_{1j} + \hat{p}_{2j}) - ln\,\hat{p}_{ij}\}^2 - \left[\sum_{j=1}^{r} \hat{p}_{ij}\{ln(\hat{p}_{1j} + \hat{p}_{2j}) - ln\,\hat{p}_{ij}\} \right]^2 \right) \quad (10)$$

$$\hat{V}(\hat{L}) = \left(\frac{\hat{L}}{N}\right)^2 \sum_{i=1}^{2} \left[\sum_{j=1}^{r} n_{ij}(\ln \hat{c}_j - \ln \hat{p}_{ij})^2 - \left\{ \sum_{j=1}^{r} n_{ij}(\ln \hat{c}_j - \ln \hat{p}_{ij}) \right\}^2 \right]. \quad (11)$$

$$\hat{V}(C\hat{X}) = \sum_{i=1}^{2} \frac{1}{N_i} \left\{ \sum_{j=1}^{r} \hat{p}_{ij}\, b_{ij}^2 - \left(\sum_{j=1}^{r} \hat{p}_{ij} b_{ij} \right)^2 \right\}, \quad (12)$$

where $b_{1j} = 2\{\hat{p}_{2j}/(\hat{p}_{1j} + \hat{p}_{2j})\}^2$ and $b_{2j} = 2\{\hat{p}_{1j}/(\hat{p}_{1j} + \hat{p}_{2j})\}^2$.

$$\hat{V}(F\hat{T}) = 1/(2\bar{n})\,(1.0 - F\hat{T}^2), \quad (13)$$

where $\bar{n} = 2N_1N_2/(N_1 + N_2)$. Formula 13 is given in van Belle & Ahmad (1974) and formulas (9) and (12) are slightly different from those in van Belle & Ahmad (1974).

Based on the above estimates of the variance and the asymptotic normality, test statistics and confidence intervals may be formed. An approximate $100(1 - \alpha)\%$ confidence interval for a measure y is

$$\hat{y} \pm z_{1-\alpha/2}\{\hat{V}(\hat{y})\}^{\frac{1}{2}}, \quad (14)$$

where $z_{1-\alpha/2}$ is the $\alpha/2$ upper percentile of the standard normal distribution. To test H_0: $y = y_0 \neq 1$, one would use

$$Z = (\hat{y} - y_0)/\{\hat{V}(\hat{y})^{\frac{1}{2}}\}. \quad (15)$$

An approximate test of $H_0: y_1 = y_2$, where y_1 is the measure at time (or location) one and y_2 is the measure at time (or location) two is

$$Z = (\hat{y}_1 - \hat{y}_2)/\{\hat{V}(\hat{y}_1) + \hat{V}(\hat{y}_2)\}^{\frac{1}{2}}, \quad (16)$$

where $\hat{V}(\hat{y}_1)$ and $\hat{V}(\hat{y}_2)$ are the respective variance estimates. Note that these results are approximate and the accuracy depends on the sample size, larger sample sizes being more accurate. If sample sizes are small (N_1 and N_2 around 20), one may use the t distribution to approximate the distribution of the test statistic with degrees of freedom determined by the estimated variances (see Snedecor & Cochran 1967, p. 115). As the normality result may be inappropriate with small sample sizes, some caution should be applied here.

One notes that the variance results in all but one case are functions of the $\hat{p}_{ij}$'s. This dependence is undesirable as most statistical procedures (such as regression) assume that the variance is the same for all data points. For the FT measure, the variance only depends on the measure and this dependence can be removed by a variance stabilizing transformation. Van Belle & Ahmad (1974) show that the distribution of

$$\hat{FT}^* = \sqrt{2}\ \text{arcsin}\ \hat{FT}$$

is asymptotically normal with variance $1/\bar{n}$. Weighted procedures (e.g. weighted regression) with weights equal to the harmonic mean sample size can then be used.

For the FT measure, an approximate 95% confidence interval is

$$\sin\,[\text{arcsin}\,(\hat{FT}) \pm z_{1-\alpha/2}(2\bar{n})^{-\frac{1}{2}}]. \quad (17)$$

This confidence interval is not symmetric. Simulation results indicate that the sampling distribution of the measure is also not symmetric so the lack of symmetry in the confidence intervals reflects the underlying distribution skewness. In Table 1, some example confidence intervals are given for data from Root (1967).

The above results apply when the overlap is not near one. As one is the upper bound of the measure, it is not reasonable for the measures to be normally distributed about the true value of one. Rather, under the hypothesis $H_0^*: y = 1$, we have

$$X_L^2 = -2N\,ln(\hat{L}), \quad (18)$$

$$X_{CX}^2 = 4\bar{n}\,(1 - \hat{C}X) \quad (19)$$

$$X_{FT}^2 = 2n\,(1 - \hat{FT}), \quad (20)$$

and $X_R^2 = 4\bar{n}\ ln(2)\,(1 - \hat{R}_0)$

Table 1. Examples of confidence intervals for data from Root (1967) on the diet of gnatcatchers (prey items are: A = Hemiptera, B = Coleoptera, C = Lepidoptera, D = Hymenoptera, E = other species).

Species	Resources					
	A	B	C	D	E	N_i
1. *Polioptila caerulea*	103	93	20	40	31	287
2. *Vireo gilvus*	23	31	132	14	13	213

Measure	Confidence intervals		
	Lower bound	Estimate	Upper bound
L	0.784	0.827	0.870
CX	0.583	0.659	0.735
FT	0.742	0.798	0.848
PS	0.381	0.449	0.518
C	0.323	0.417	0.510
R_0	0.653	0.722	0.791

are asymptotically chi-square with $r-1$ degrees of freedom [the results for the CX and FT measures are explicitly derived in van Belle & Ahmad (1974) and result for L is easily seen by recognizing that X_L^2 is the likelihood ratio statistic]. A similar result holds for Horn's information measure as it is related to Petraitis' likelihood measure (Petraitis 1979). X_L^2, X_{CX}^2 and X_{FT}^2 are two sample versions of the goodness-of-fit statistics considered in Bishop et al. (1975, p. 513). Van Belle & Ahmad (1974) show that Morisita's measure (C_1) is asymptotically noncentral chi-square. This distribution depends on the resource use distributions and is therefore difficult to table. We have not been able to find the asymptotic distribution of the PS measure under H_0^*. Simulation methods could be used in this case to test the hypothesis by taking multinomial samples of size N_1 and N_2 from the vector $\underset{\sim}{\hat{c}} = (\hat{c}_1, \hat{c}_2, \ldots, \hat{c}_r)$, where $\hat{c}_j$ is the estimate of the common usage ($\hat{c}_j = (n_{1j} + n_{2j})/(N_1 + N_2)$). The estimate of PS is then computed. This procedure is repeated, say, 99 times, and the observed estimate is compared with the fifth smallest (for a 95% test). If the observed value is less, the hypothesis is rejected.

Petraitis (1979) suggests that since X_L^2 is asymptotically chi-square, one may compare two measures by forming the ratio of transformed measures and treat the result as a random variable from an F distribution. However, note that the central chi-square assumption is only valid under the hypothesis H_0^*: L = 1. Hence the ratio will in general have a doubly noncentral F distribution. The approach using the delta method is appropriate.

Simulation results

To assess the accuracy of the variance approximations and assess normality, a simulation study was undertaken. Samples of size N_1 and N_2 were taken from a $M(N_1, \underset{\sim}{p}_1)$ and $M(N_2, \underset{\sim}{p}_2)$ and each measure computed. The variance was computed based on 2000 repetitions. To judge asymptotic normality, rankit plots (Weisberg 1981) were drawn based on 200 sample values. We present some results here. Additional results are given in Smith (1982).

Comparison of the simulated variances with the delta estimates are given in Table 2. In general, the

Table 2. Comparison of simulated variance with variance estimated using the delta method. $N_1 = N_2 = 50$, $\underset{\sim}{p}_1 = (0.4, 0.3, 0.1, 0.05, 0.05)$ and $\underset{\sim}{p}_2$ is a permutation of $\underset{\sim}{p}_1$.

	permutation 1	2	3
L	0.775	0.923	0.983
Estimated variance	0.0021	0.0011	0.0003
Simulated variance	0.0023	0.0013	0.0006
CX	0.549	0.852	0.966
Estimated variance	0.0083	0.0039	0.0012
Simulated variance	0.0085	0.0041	0.0017
FT*	1.15	1.64	1.96
Estimated variance	0.020	0.020	0.020
Simulated variance	0.028	0.023	0.020
R_0	0.634	0.885	0.975
Estimated variance	0.0072	0.0028	0.0007
Simulated variance	0.0085	0.0034	0.0014
C_1	0.327	0.809	0.982
Estimated variance	0.0083	0.0081	0.0004
Simulated variance	0.0081	0.0078	0.0018
PS	0.400	0.750	0.900
Estimated variance	0.0044	0.0038	0.0029
Simulated variance	0.0050	0.0049	0.0039

method gives accurate results although the variance tends to be underestimated. As the measures near 1, the accuracy becomes worse and the delta method estimates the variance to be zero when p_1 and p_2 are the same. Although the variance for Petraitis' measure (L) appears to be considerably less than the others, this is due to the difference in range. As the minimum of L is 0.5, the variance should be multiplied by 4.0 for purposes of comparing variances.

Some indication of the normality of the estimators is given in Figure 1, 200 ordered values of the measures are plotted against 200 expected order statistics from a N(0,1) distribution. Although there may be concern over the details, the results are, in general, encouraging. The sample sizes in Figure 1 are small compared to most studies. Some additional examples to support the use of normality are given in Smith (1982). Also note in Figure 1 that the bias may be large with small sample sizes. The problem of bias is considered elsewhere (Smith & Zaret 1982).

Discussion to part 1

Petraitis (1979) first recognized that the overlap measures commonly used could not be tested statistically. He suggested using a measure based on the likelihood ratio statistic. We have proposed two additional measures based on alternate test statistics, the chi-square goodness-of-fit statistic and Matusita's statistic. All three measures yield similar results when applied to data. Confidence intervals are available for these measures and also for some more commonly used overlap measures. The best results are for the FT measure as the variance (and, hence, confidence interval) does not depend on the unknown resource usages but rather on the measure itself. Further, a transformation is available to

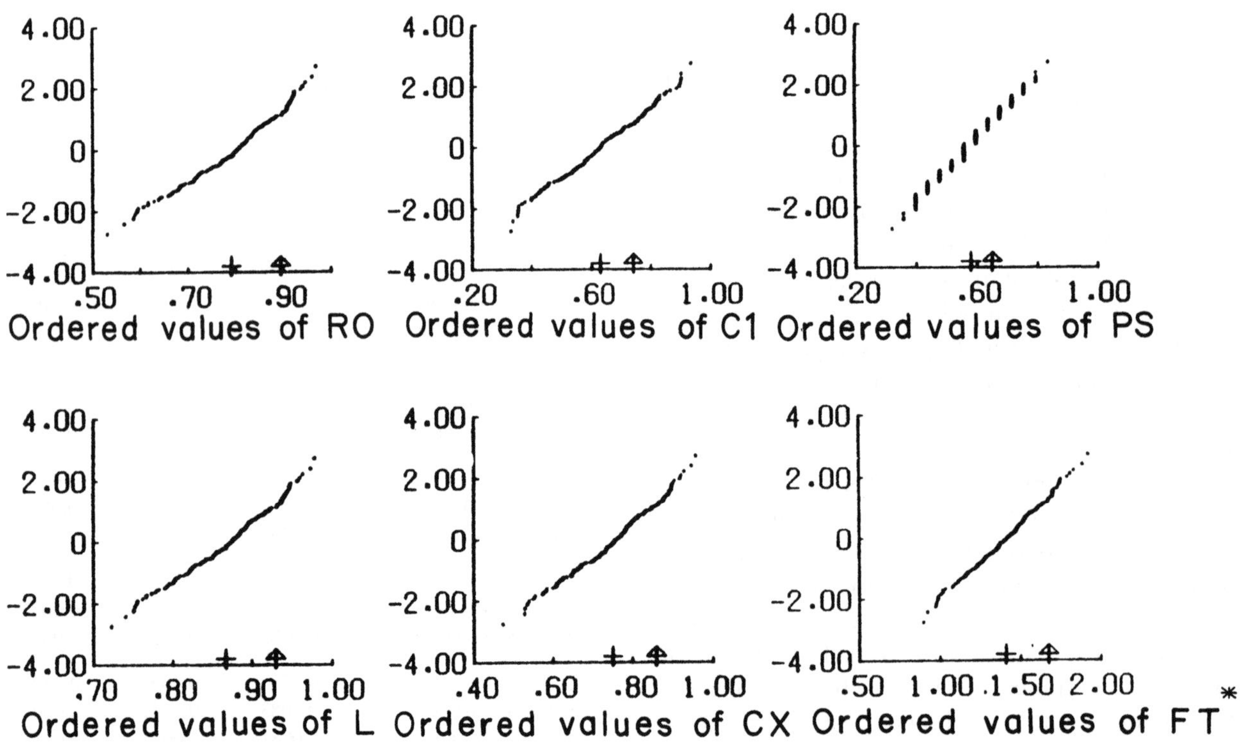

Fig. 1. Rankit plots for six overlap measures. The y values are given by $y_i = \Phi^{-1}$ (i-3/8)/(n+1/4) and x_i is the ith sample order statistic for the overlap measures. Based on 200 simulated overlap measures with $N_1 = N_2 = 25$, $\underset{\sim}{p}_1$ = (.3, .2, .15, .10, .10, .10, .05) and $\underset{\sim}{p}_2$ = (.15, .10, .05, .30, .2, .1, .1) The symbol ↑ refers to the actual value and + the mean simulated value.

remove this dependence and make the variance depend only on the sample sizes (N_1 and N_2). Hence, this measure is recommended for use with statistical procedures.

Some caution must be used in applying these results. First, if one computes confidence intervals on all possible pairs of species, simultaneous methods should be used (see Miller 1966). Second, the methods are not valid for comparing overlaps within a sample (e.g. species 1 and 2 with species 1 and 3) as the measures are not independent.

All methods used here in deriving test statistics and confidence intervals assume a multinomial model. This model is based on an assumption of random use of resources and an assumption that the species do not affect the resources (the p_{ij}'s do not change). In many cases, the resources are patchy and hence the multinomial model may not apply. Additionally, predators may affect the proportions of prey available if the predators are selective. Hence the estimates of variance are likely to be lower bounds on the true variances and are estimates for a fixed time point. One approach incorporating patchiness is to model the patchiness and compare models over species (see Sandusky & Horne 1978 for examples). Alternately, when there are multiple observations (e.g., individual stomach contents) a nonparametric approach using the jackknife method is useful (see Smith et al. 1979). We note however, that the multinomial model has been applied successfully in some studies of resource use (Fienberg 1970, Schoener 1970).

Overlap measures have been criticized for not considering resource availability in the measure of overlap (Lawlor 1980). Resources could be included by using resource availability as weights in the computation of the measure. I R_j denotes the availability of resource j, then the overlap measure can be computed based on the proportion $p_{ij}^* = p_{ij}/R_j$. The new measure would, however, not have a maximum of one. To alleviate this problem, one would normalize the proportion by dividing by the sum of the weights ($\sum 1/R_j$). Confidence intervals could then be estimated using the delta method on the new measures, the simplest approach assuming the resource availabilities as fixed. In this case, the chi-square tests based on transformation of the adjusted measures have noncentral chi-square distributions with the noncentrality parameter depending on the R_j's.

Table 3 shows an application of the goodness-of-fit statistics X_L^2, X_{CX}^2, and X_{FT}^2 to the data of Zaret & Rand (1971), in order to test the null hypothesis that all pairs of species are similar in their resource use. Note, however, that using the preferred FT measure, the null hypothesis is rejected for all but one of the possible combinations of two species (i.e., $X_{FT}^2 > x_{\mathrm{df}\,0.05}^2$). Only in two cases, using overlap measures L and CX for species pair 2, 5 (*Neoheterandria* and *Gephyrocharax*), do we accept the null hypothesis of complete overlap. It is rarely possible to assert that two species' diets show complete overlap.

Conclusion

Our goal in this part was to develop statistical means of evaluating differences in the diets of species A and B. Using our approach, one can (a) compare a calculated overlap measure with a specified value, e.g., does it differ from 1.0, 0.5, and (b) compare overlap values for the same species pair at different times of the day, e.g., day versus night, spring versus summer. Note, however, that the rejection of the hypothesis of complete overlap does not necessarily mean that the overlap is not biologically significant. The choice of the correct hypothesis is necessary to infer biological significance.

Part 2: Morphological diversity

Large numbers of ecologically similar and systematically related species occur together in the Amazon Basin. As an obvious example, the species of characin fishes are among the most diverse groups of vertebrates anywhere in the world (Géry 1978). A large amount of ecological theory concerns such associations of related similar species (Van Valen 1965, MacArthur & Levins 1967, Roughgarden & Feldman 1975), and there is also empirical work on such closely related groups as Darwin's finches (Smith et al. 1978 for a good review). In cases of extensive dietary overlap, character displacement is

Table 3. Comparison of test statistics for data from Zaret & Rand (1971). Note that critical values are adjusted for non-shared resources.

Genera	Resources 1	2	3	4	5	6	7	8	9	10	11	12	N_i
1. *Astyanax*	9	12	7	3	21	0	0	0	11	0	0	0	63
2. *Neoheterandria*	0	6	0	0	7	3	0	0	0	1	0	0	17
3. *Aequidens*	6	1	0	0	29	0	0	4	0	0	17	0	57
4. *Reoboides*	0	6	3	0	23	0	1	10	0	0	0	4	47
5. *Gephyrocharax*	0	9	3	6	15	2	0	0	0	0	0	0	35

Test statistics

Species pair	X_L^2	X_{cx}^2	X_{FT}^2	df	$\chi^2_{dF,0.05}$
1.2	28.36	20.57	39.65	8	15.51
1.3	70.78	53.13	97.61	8	15.51
1.4	54.11	39.78	76.65	9	16.92
1.5	26.48	21.13	38.30	7	14.07
2.3	38.55	26.78	48.55	8	15.51
2.4	24.86	18.24	33.82	9	16.92
2.5	11.28	8.97	16.32	7	14.07
3.4	49.33	37.58	68.32	8	15.51
3.5	59.26	48.54	80.82	8	15.51
4.5	32.42	24.01	46.04	9	16.92

the process usually invoked as being expected (Slatkin 1976). With closely related species, morphological measurements have been used as indices of competition (Cody 1974). Some of these assumptions are currently the subject of intense debate (e.g. Conner & Simberloff 1979, Strong et al. 1979, Grant & Abbott 1980, Diamond & Gilpin 1982).

We propose here a different way of looking at morphological diversity and its origin, echoing Connell (1980). Consider the following simple scenario. Let us suppose a hypothetical continent with a single generalized, opportunistically feeding characoid species covering the entire land mass (see Fig. 2). Now suppose that some geologic event fragments this single species into three genetically isolated populations; this may be thought of as a 'vicariant' event (term adopted from Croizat 1964). During this period of isolation, we may assume that the three populations experience somewhat different environments, and yet continue to feed opportunistically. Thus population A_1 may find itself at higher altitudes or in a rain shadow, the population density of A_2 may increase, the vegetation of the area occupied by population A_3 may become extinct. As natural selection, sexual selection, intraspecific competition, for example, proceed, we may see allopatric speciation occur, resulting in three species, namely, A_1, A_2 and A_3. Finally, then, let us allow these three species to become sympatric, as the barriers that originally subdivided the populations disappear or become ineffective due perhaps to stream capture.

It is expected that these three species will differ somewhat in their feeding preferences, habitats, behaviors, etc. It is quite possible that these three species, when studied by an ecologist, will seem to partition resources efficiently (as in MacArthur 1958, Werner 1977). However, we have just shown that any observed resource partitioning by these species cannot be considered prima-facie evidence

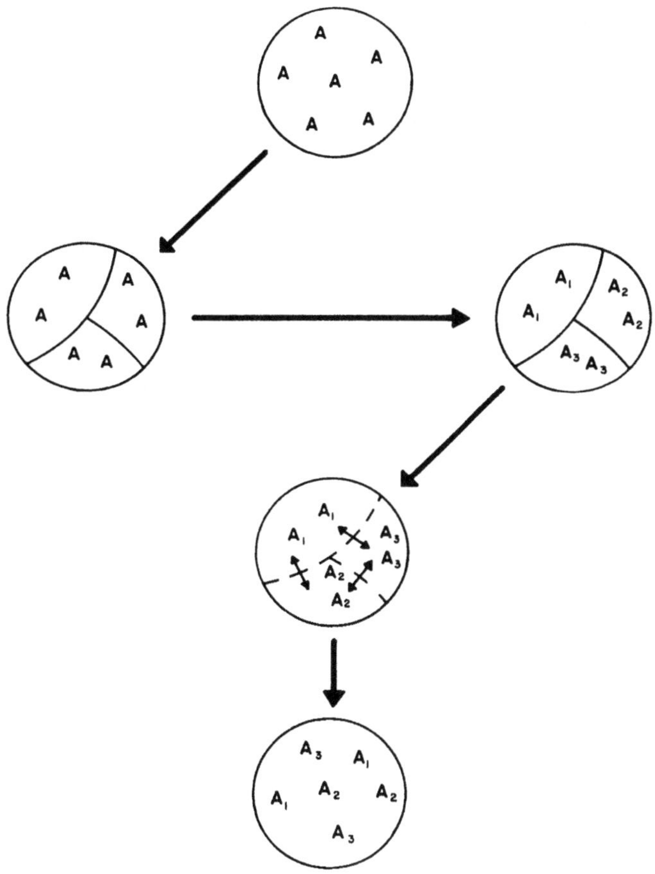

Fig. 2. Schematic diagram illustrating an alternative hypothesis for the origin of morphological diversity. See text for explanation.

of character displacement or any other manifestation of interspecific competition, the most commonly used explanations for interspecific differences in resource use. On the contrary, this phenomenon is more likely a normal consequence of allopatric speciation.

As an example, let us consider studies which have been done on closely related fish species, such as work of Earl Werner and his associates, examining the role of fish interspecific competition in Michigan lakes (Werner 1974, Werner & Hall 1976, Werner 1977). The assumption has been that the differences between say, *Lepomis macrochirus* and *Lepomis cyanellus*, are due to interspecific competition. Yet this mechanism cannot be distinguished from the alternative hypothesis presented in Figure 2. What Werner and his associates describe as a result of interspecific competition, can just as easily be explained as the consequence of three previously developed species (A_1, A_2, A_3) engaged in biological accommodation after becoming sympatric. This scenario can be accomplished without having to invoke the process of interspecific competition.

If one accepts the model as plausible, the question of time becomes critical in being able to evaluate which of these hypotheses is correct. Otherwise ecologists may be treating an adaptive response to short-term variation in the animal's environment, as analogous to an evolutionary response to long-term change which occurred over millions of years. In this model, niches become a way of describing the present relationships among species and say nothing about the origin of species. The question of time becomes the critical one, and the most important goal is finding ways of determining whether species living together have evolved together, in which character displacement becomes an issue, or whether given species have different traits as a result of allopatry and secondary adaptation. Without being able to distinguish between these two hypotheses, one cannot interpret the origin of similar morphologies among closely related species.

Acknowledgments

We thank many colleagues for help and inspiration, including Mark Anderson, John Endler, Bill Fink, Huey Lin, E. Lok, R.T. Paine, Jeanette Pederson, Barbara C. Peterson, Gordon Swartzman, and Dan Brooks for the companionship on board the ship 'Eastward', that finally germinated in this paper. Financial support came from NSF Grant BSR 80-03203.

References cited

Abrams, P. 1980. Some comments on measuring niche overlap. Ecology 61: 44–49.

Bishop, Y.M.M., S.E. Feinberg & P.W. Holland. 1975. Discrete multivariate analysis: theory and practice. The MIT Press, Cambridge. 557 pp.

Cody, M.L. 1974. Competition and the structure of bird communities. Princeton University Press, Princeton. 318 pp.

Connell, J.H. 1980. Diversity and the coevolution of competitors, or the ghost of competition past. Oikos 35: 131–138.

Connor, E.F. & D.S. Simberloff. 1979. The assembly of species communities: chance or competition? Ecology 60: 1132–1140.

Croizat, L.C.M. 1964. Space, time, form: the biological synthesis. Published by the author, Caracas, 881 pp.

Diamond, J.M. & M.E. Gilpin. 1982. Examination of the 'null' model of Connor and Simberloff for species co-occurrences on islands. Oecologia 52: 64–74.

Fienberg, S.E. 1970. The analysis of multidimensional contingency tables. Ecology 51: 419–432.

Fisher, R.A. 1958. Statistical methods for research workers. 13th ed., revised. Hafner Publishing Co., New York. 362 pp.

Fryer, G. & T.D. Iles. 1972. The cichlid fishes of the great lakes of Africa. T.F.H. Publications, Neptune City. 641 pp.

Géry, J. 1978. Characoids of the world. T.F.H. Publications, Neptune City. 672 pp.

Goodman, L.A. & W.H. Kruskal. 1979. Measures of association for cross classifications. Springer-Verlag, New York. 146 pp.

Goulding, M. 1980. The fishes and the forest; explorations in Amazonian natural history. University of California Press, Berkeley. 280 pp.

Grant, P.R. & I. Abbott. 1980. Interspecific competition, island biogeography and null hypotheses. Evolution 34: 332–341.

Hall, D.J., W.E. Cooper & E.E. Werner. 1970. An experimental approach to the production dynamics and structure of freshwater animal communities. Limnol. Oceanogr. 15: 829–928.

Horn, H.S. 1966. Measurement of overlap in comparative ecological studies. Amer. Nat. 100: 419–423.

Hurlbert, S.H. 1978. The measurement of niche overlap and some relatives. Ecology 59: 67–77.

Lawlor, L.P. 1980. Overlap, similarity and competition coefficients. Ecology 61: 245–251.

Linton, L.R., R.W. Davies & F.W. Wrona. 1981. Resource utilization indices: an assessment. J. An. Ecol. 50: 283–292.

MacArthur, R.H. 1958. Population ecology of some warblers of northern coniferous forests. Ecology 39: 599–610.

MacArthur, R.H. 1968. The theory of the niche. pp. 159–176. *In*: R.C. Lewontin (ed). Population Biology and Evolution, Syracuse University Press, Syracuse.

MacArthur, R.H. & R. Levins. 1967. The limiting similarity, convergence and divergence of coexisting species. Amer. Nat. 101: 377–385.

Matusita, K. 1955. Decision rules based on distance, for problems of fit, two samples and estimation. Ann. Math. Stat. 26: 631–640.

Miller, Rupert G. 1981. Simultaneous statistical inference. 2nd ed., Springer-Verlag, New York. 299 pp.

Morisita, M. 1959. Measuring interspecific association and similarity between communities. Memoirs of the Faculty of Science of Kyushu University, Series E, Biology 3: 64–80.

Petraitis, P.S. 1979. Likelihood measures of niche breadth and overlap. Ecology 60: 703–710.

Renkonen, O. 1938. Statistisch-ökologische Untersuchungen über die terrestrische der finnischen Bruchmoore. Annales Zoologici Societatis Zool. Bot. Fennicae, Vanamo. 231 pp.

Ricklefs, R.E. & M. Lau. 1980. Bias and dispersion of overlap indices: results of some Monte Carlo simulations. Ecology 61: 1019–1024.

Root, R.B. 1967. The niche exploitation pattern of the blue-gray gnatcatcher. Ecological Monographs 37: 317–350.

Roughgarden, J. & M. Feldman. 1975. Species packing and predation pressure. Ecology 56: 489–492.

Samuel-Cahn, E. 1975. Remark on a formula by Fisher. Journal of the American Statistical Association 70: 720.

Sandusky, J.C. & A.J. Horne. 1978. A pattern analysis of Clear Lake phytoplankton. Limnol. Oceanogr. 23: 636–648.

Schoener, T.W. 1970. Non-synchronous spatial overlap of lizards in patchy habitats. Ecology 51: 408–418.

Schoener, T.W. 1974. Resource partitioning in ecological communities. Science 185: 27–39.

Seber, G.A.F. 1973. The estimation of animal abundance and related parameters. Charles Griffin, London. 654 pp.

Slatkin, M. 1980. Ecological character displacement. Ecology

61: 163–177.

Smith, E.P. 1982. The statistical properties of biological indices of water quality. Ph.D. Thesis, University of Washington, Seattle, 229 pp.

Smith, E.P. 1983. Niche breadth, resource availability and inference. Ecology (in print).

Smith, E.P. & T.M. Zaret. 1982. Bias in estimating niche overlap. Ecology 63: 1248–1253.

Smith, J.N.M., P.R. Grant, B.R. Grant, I.J. Abbott & L.K. Abbott. 1979. Seasonal variation in feeding habits of Darwin's ground finches. Ecology 59: 1137–1150.

Smith, W., D. Kravitz & J.F. Grassle. 1979. Confidence intervals for similarity measures using the two sample jacknife. pp. 253–262. *In*: L. Orloci, C.R. Rao & W.M. Stiteler (ed) Multivariate Methods in Ecological Work, International Cooperative Publishing House, Fairland.

Snedecor,G.W. & W.G. Cochran. 1967. Statistical methods. Sixth Edition. Iowa State University Press, Ames. 507 pp.

Strong, D.R., L.A. Szyska & D.S. Simberloff. 1979. Tests of community-wide character displacement against null hypotheses. Evolution 33: 897–913.

Van Belle, G. & I. Ahmad. 1974. Measuring affinity of distributions pp. 651–668. *In*: F. Proschan & R.J. Serfling (ed.) Reliability and Biometry, SIAM Publications, Philadelphia.

Van Valen, L. 1965. Morphological variation and width of the ecological niche. Amer. Nat. 94: 377–390.

Weisberg, S. 1981. Applied linear regression. John Wiley & Sons, New York. 283 pp.

Werner, E.E. 1974. The fish size, prey size, handling time relation in several sunfishes and some implications. J. Fish. Res. Board Can. 31: 1531–1536.

Werner, E.E. 1977. Species packing and niche complementarity in three sunfishes. Amer. Nat. 111: 553–578.

Werner, E.E. & D.J. Hall. 1976. Niche shifts in sunfishes: experimental evidence and significance. Science 191: 404–406.

Whittaker, R.H. & S.A. Levin (ed). 1975. Niche: theory and application. Dowden, Hutchinson & Ross, Stroudsberg. 448 pp.

Zaret, T.M. & A.S. Rand. 1971. Competition in tropical stream fishes: support for the competitive exclusion principle. Ecology 52: 336–342.

Salminus affinis (top fish) and *Ichthyoelephas longirostris* from Steindachner (1880), Zur Fisch-Fauna des Cauca und der Flüsse bei Guayaquil. Denkschr. Akad. Wiss. Wien 42: 55–104.

The status of studies on South American freshwater food fishes

Rosemary H. Lowe-McConnell
Streatwick, Streat via Hassocks, Sussex, England

Keywords: Tropical fishery, Rivers, Standing stock, Catch, Production, Ichthyomass, Conservation, Floodplains, Market statistics

Synopsis

This paper examines the main freshwater food fishes in South America based on their biology and market statistics. In South America freshwater fisheries are based on riverine fishes, predominantly otophysan characoids and siluroids, with osteoglossids, cichlids, sciaenids and clupeids locally important. Some seventy species, or species groups, are marketed; many of the large species are widely distributed, from the Rio Orinoco to the Rio Paraná. The limited studies from tropical freshwaters of Africa, Asia and South America are compared for quantitative data on standing stocks, yields and biological production. Declines in catches of preferred species near large centers of human population suggest local overexploitation, but catch statistics are insufficiently documented to prove this. Riverside deforestation may present a greater threat to fish stock decline than overexploitation.

Introduction

South America has the richest freshwater fish fauna of any zoogeographical region, but the rapidly increasing human population means that there is a steeply rising demand for fish as food. Habitats are being altered at an accelerating rate by deforestation, agriculture, industrialization, pollution and most recently the building of hydroelectric dams, resulting in actual or potential depletion of stocks of some preferred food fishes. This paper focuses on the kinds of information needed to develop fisheries, and how far studies have progressed toward obtaining the necessary data.

The rich freshwater fish fauna spread over such a vast and relatively unexplored continent is a difficult one to study. Böhlke et al. (1978) suggest that perhaps 30 to 40% of the South American fish fauna still remains undescribed. Although nearly all of the larger food fishes are well known, many of the original descriptions are inadequate, and food fish specimens are still poorly represented in museums. Most groups are in need of taxonomic revision. The difficulties of obtaining the older literature and of examining type material that is widely scattered in museums in Europe and North America impede such studies. The large size of many of the species makes them difficult to collect, and their long migratory movements in the vast and complex river systems make them more difficult to study than lacustrine or stream-dwelling fishes. The information available on all aspects of fish biology is fragmentary, widely scattered in journals and reports, reflecting the distribution of the fishery scientists rather than the fishes. The major river systems of South America connect with one another or have done so in the recent past; many of the large fishes are widely distributed throughout the con-

Thomas M. Zaret (ed.), Evolutionary ecology of neotropical freshwater fishes. ISBN 90 6193 823 6

tinent (Orinoco to Paraná systems) where comparable habitats exist. Studies in one area are therefore relevant to other areas.

The fisheries

Sources of information

Data on South American food fishes come from analyses of market statistics and from biological observations made around the various centers indicated in Figure 1. Brazilian market statistics showed that in 1974, 15% (114,189 t) of the total (737,019 t) fish handled came from freshwaters, 67% of this freshwater catch from the Amazon region (Smith 1981). The species caught from the various Amazon tributaries (monthly totals 1969–1974) are given by Honda et al. (1975) and analyzed by Petrere (1982). Throughout the continent, however, freshwater fishes are particularly important for subsistence fishermen whose catches do not appear in market statistics. Research has been particularly active in the Central Amazon basin around Manaus, where statistics of catches from a wide area are being examined and autecological studies of the main food fishes are in progress. Other areas for Table 1 include the Rupununi area of Guyana, the Orinoco system in Venezuela, various parts of the Paraná-Paraguai system, the Mogi Guaçu tributary of the Upper Paraná in Brazil and the Middle Paraná in Argentina.

The main food fish species

There is no one checklist or identification manual for South American freshwater food fishes and the literature is scattered. Regional bibliographies and sources for identifying particular fish groups are listed by Lowe-McConnell and Howes (1981). Between 2500 and 3000 species have already been described from South America, over 85% of them otophysan fishes, characoids and siluroids (catfishes) represented by about equal numbers of species. These include most of the large food fishes that are often migratory, schooling, and easy to catch in large numbers. Many other species are small; some are of commercial importance as aquarium fishes. Of the twelve characoid families present (as recognized by Greenwood et al. 1966) only six are regularly represented in fish markets: Characidae, Prochilodontidae, Curimatidae, Hemiodontidae, Anostomidae, Erythrinidae; of the fourteen catfish families only five: Pimelodidae, Doradidae,

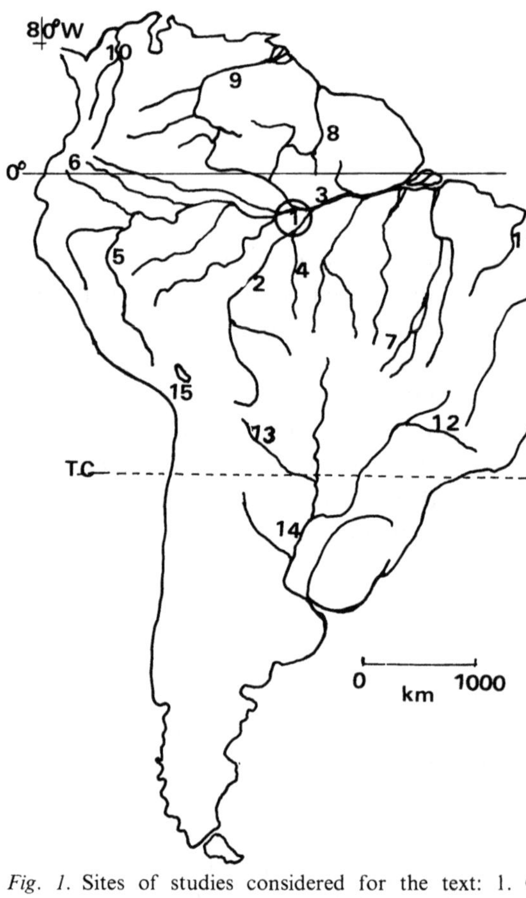

Fig. 1. Sites of studies considered for the text: 1. Central Amazon (Manaus): Mainly from I.N.P.A., National Institute of Amazon Research (in Port.), Manaus, Brazil; 2. Rio Madeira: Goulding 1980, 1981; 3. Itacoatiara: Smith 1981; 4. Aripuaná: M.G.M. Soares 1979; 5. Peruvian Amazon: Chapman 1978; 6. Ecuador: Saul 1975; 7. Rio das Mortes: Lowe-McConnell (unpublished); 8. Guyana: Lowe-McConnell 1962, 1969, 1975; 9. Orinoco: Mago-Leccia 1970, 1972, Novoa & Ramos 1978; 10. Magdalena: Kapetsky 1977, Miles 1947, 1973; 11. Rio Parnaíba: Paiva 1973, Paiva & Gesteira 1977; 12. Mogi Guaçu: Godoy 1967, 1975; 13. Pilcomayo: Bayley 1973; 14. Paraná: Bonetto et al. 1969, Bonetto 1975, Cordiviola de Yuan & Pignalberi 1981, Ringuelet et al. 1967; 15. L. Titicaca: Eigenmann & Allen 1942.

Hypophthalmidae, Loricariidae, Callichthyidae. With the exception of the Characidae (also found in Africa) and the Ariidae (widely distributed marine catfishes), these twenty-six families are all endemic to the neotropics.

Subsistence fishermen use a wider spectrum of fishes, including acestrorhynchine characoids and auchenipterid catfishes. Species of two other families, Cetopsidae and Trichomycteridae, small predatory or parasitic catfishes, are also of commercial significance as they damage other catfishes when they have been caught. The third most speciose group of South American freshwater fishes, the electrogenic gymnotoids, are rarely found in fish markets. These are mostly small species, and there are taboos against eating gymnotoids in some areas. The cichlids, which have speciated so abundantly in African lakes where they are the main food fishes, make up only ~2% of the South American fauna. They are easy to catch on hooks, in traps, using small nets, and are important in subsistence fisheries, but only the larger species appear in fish markets. Of the archaic fishes, the osteoglossids *Arapaima* and *Osteoglossum* are popular food fishes wherever these occur. Of the ~50 freshwater species representing families of marine fishes, the sciaenid *Plagioscion* species and clupeid *Pellona* species are common in certain fish markets.

The main food fishes in various river systems are listed in systematic order in Table 1, the names culled from the sources indicated in Figure 1. Local names are also given, as market statistics are collected under local names. Moreover, in the present state of knowledge a local name may cover more than one species e.g. pacú for the deep-bodied characoids).

Species composition

Amazonas. — Manaus market statistics given here for 1976 are of fishes brought to Manaus from a very wide area (listed and mapped by Petrere 1978a, 1978b, who has since completed comparable analyses for later years, Petrere 1982). Manaus catches appear to have declined by about one-half since Petrere's 1976 data (Goulding personal communication). These 1976 catches came from the Brazilian border over 1700 km upriver to 600 km downriver from Manaus, as well as from long distances up the main tributary rivers (to São Paulo de Olivença 1508 km on the Solimões, 870 km on the Japurá, 1719 km on the Juruá, 1409 km on the Jutaí, 1360 km on the Purus, to Manicoré on the Rio Madeira 619 km from Manaus, and to Barcelos 466 km up the Rio Negro). A Manaus boat might be away for 40 days, for example, fishing for tambaqui in 103 different places, using one of eight fishing methods. Nearly half the fish (48%) came from openwater seines (arrastao). These included a wide spectrum of species, particularly migratory species such as prochilodontids (21% curimata, 20% jaraqui) and serrasalmines (16% tambaqui, 9% pirapitinga, 10% pacu). Some 34% came from gillnets (in which the catch was 95% tambaqui, mainly from flooded forests); 9% came from shore seines (arrastadeira) also used to catch migrating fishes (including 63% jaraqui, 19% tambaqui, 10% matrinchão); 3% came from tridents used with lamps at night (taking 95% tucunare, *Cichla*); 2% from trot-lines used for tambaqui (catch 96% tambaqui), 1% from baited lines (which took 67% tucunare and 16% other cichlids); 0.5% from handlines (catch 96% pescada, *Plagioscion*); 0.3% from harpoons for *Arapaima* (catch 87% *Arapaima*).

The total 30,929 t landed at Manaus in 1976 included over 30 species, or species groups, of fish. *Colossoma* and *Mylossoma* were the main genera, followed by prochilodontids, then other relatively large characoids such as anostomids and *Brycon*. Some smaller genera, *Triportheus* and curimatids, were also important. The only catfishes recorded here were small numbers of loricariids, as the larger pimelodid and doradid catfishes are not kept when preferred species are available. Of the non-otophysan fishes, *Cichla*, *Osteoglossum* and *Plagioscion* were here the most important.

Data from the Porto Velho market ~ 900 km up the Rio Madeira, and information on the natural history of the migratory characoids and catfishes of the Rio Madeira, have been analyzed by Goulding (1979, 1980, 1981). Here the fishes came mainly either from (1) seine catches in rivers, (2) the Teotônio cataracts some 25 km above Porto Velho, where fishermen gaffed and netted large pimelodids

Table 1. The main food fish species in various South American rivers.

Central Amazon	Orinoco	Guyana	Mogi Guaçu	Paraná
Engraulidae				*Lycengraulis olidus*
				anchoa de rio
Clupeidae				
Pellona castelnaeana	*Pellona* sp.			
apapá amarelo	sardinata			
Osteoglossidae				
Arapaima gigas		*Arapaima gigas*		
pirarucu		arapaima		
Osteoglossum bicirrhosum		*Osteoglossum bicirrhosum*		
aruaná		arawana		
Characidae				
Brycon sp.	*Brycon* sp.		*Brycon lundii*	*Brycon orbignyanus*
matrinchão	bocona, palambra		piracanjuba	pirapita
Brycon cephalus				
jatuarana			*Salminus maxillosus*	*Salminus maxillosus*
Triportheus elongatus			dorado	dorado
sardinha comprida			*Salminus hilarii*	
Triportheus angulatus			tabarana	
sardinha chata				
Triportheus rotundatus				
sardinha				
Colossoma macropomum	*Colossoma macropomum*			*Colossoma mitrei*
tambaqui	cachama			pacu caranha
Colossoma bidens	*Colossoma brachypomum*	*Colossoma brachypomum*		
pirapitinga	morocoto	morocut		
Mylossoma albiscopus	*Mylossoma* sp.			
pacú vermelho	palometa			
Mylossoma duriventris				
pacú toba				
Mylossoma aureum				
pacú branco				
Myleus spp.		*Myleus pacu*	*Myleus asterias*	
pacú mafura		red pacú	pacú	
Metynnis spp.		*Metynnis* spp.		
pacú		pacú		
Serrasalmus rhombeus	*Serrasalmus* spp. (3)	*Serrasalmus rhombeus*		
piranha branca, preta	caribe	black pirai		
Serrasalmus nattereri	*Serrasalmus nattereri*	*Serrasalmus nattereri*		
piranha cajú	caribe colorao	red pirai		

Erythrinidae				
Hoplias malabaricus	*H. malabaricus*	*H. malabaricus*		*Hoplias malabaricus*
traira	guabina	huri		tararira
	Hoplias macrophthalamus	*H. macrophthalmus*		
	aimara	haimara		
Hoplerythrinus unitaeniatus	*H. unitaeniatus*	*H. unitaeniatus*		
jiju	guabina	yarrow		
Cynodontidae				
Hydrolycus spp.	*Hydrolycus* spp.	*Hydrolycus scomberoides*		
peixe cachorro	payara	baiara		
Prochilodontidae				
Prochilodus nigricans	*Prochilodus mariae*	*Prochilodus rubrotaeniatus*	*Prochilodus scrofa*	*Prochilodus platensis*
curimatá	coporo	yakutu	curimbatá	sábalo
Semaprochilodus taeniurus	*Semaprochilodus laticeps*	*Semaprochilodus insignis*		
jaraqui escama fina	sapoara	jarake		
Semaprochilodus theraponura	*Semaprochilodus kneri*			
jaraqui escama grossa	bocachico			
Semaprochilodus laticeps				
jaraqui açu				
Curimatidae				
Curimatus spp.	*Curimatus* sp.			
branquinha	blanquita			
Anostomidae				
Leporinus friderici	*Leporinus* spp.	*Leporinus friderici*	*Leporinus elongatus*	*Leporinus* spp.
aracu de cabeça gorda	boquimi, mije	komona	piapara	bogas
Leporinus fasciatus		*Leporimus fasciatus*	*Leporinus copelandii*	*Leporinus obtusidens*
aracu paca		kibihee	piava	boga
			Leporinus octofasciatus	
			piava ussu	
Schizodon fasciatum	*Schizodon isognathus*	*Schizodon fasciatum*	*Schizodon nasutus*	*Schizodon fasciatum*
aracu pintado, comúm	pijotero	kwan	taguara	boga lisa
Rhytiodus spp.			*Leporellus vittatus*	
aracu pau de negro, piao			solteira	
Hemiodontidae				
Hemiodus microlepis				
orana				
Anodus sp.				
orana				

Table 1. (Continued).

Central Amazon	Orinoco	Guyana	Mogi Guaçu	Paraná
Pimelodidae				
Brachyplatystoma flavicans	*Brachyplatystoma* sp.			
dourada	dorado			
Brachyplatystoma vaillantii	*B. vaillantii*	*B. vaillantii*		
piramutaba, piaba	lau-lau	lau-lau		
Brachyplatystoma filamentosum	*B. filamentosum*	*B. filamentosum*		*B. filamentosum*
piraiba, filhote		kumakuma, manari		bagre branco, piraiba
Pseudoplatystoma fasciatum	*P. fasciatum*	*P. fasciatum*		*P. fasciatum*
surubím	rayao	tiger fish, kurutu		surubi, sorubim
Pseudoplatystoma tigrinum	*P. tigrinum*			*P. coruscans*
caparari	rayao			surubi manchado
Pinirampus pirinampu	*P. pirinampu*			
piranambu, barba chata	blanco pobre			
Phractocephalus hemiliopterus	*P. hemiliopterus*	*P. hemiliopterus*		
pirarara	bagre cajaro	aruarina		
Paulicea lutkeni				*Paulicea lutkeni*
jaú				manguruyu
Sorubim lima				
bico do pato				
Sorubimichthys planiceps	*S. planiceps*			*Sorubimichthys planiceps*
peixe lenha	doncella			pirayape ani
Merodontotus tigrinus	*Sorubimichthys* sp.			
dourada fita	cabo de hacha			
Hemisorubim platyrhynchos	*H. platyrhynchos*			*H. platyrhynchos*
bagre braço de moca	bagre cupido			jiripoca
Goslinia platynema	*G. platynema*			
babão	bagre garbanzo			
Callophysus macropterus	*C. macropterus*			
piracatinga, pintadinho	mapurite			
Pimelodus blochii	*Pimelodus clarias*	*P. clarias*	*P. clarias*	*Pimelodus maculatus*
mandi	mandi	mandi	mandi amarelo	bagre amarillo
	Leiarius marmoratus			*Pimelodus albicans*
	bagre yaque			bagre blanco
	Pseudopimelodus sp.			*Luciopimelodus pati*
	bagre amarillo			pati, piracatinga
Doradidae				
Megalodoras irwini	doradids			*Rhinodoras d'orbignyi*
bacu rebeca	guitarrilla, sierra			armado amarillo
Pterodoras granulosus				*Pterodoras granulosus*
bacu comun				armado comun

Lithodoras dorsalis				
pacu pedra				
Oxydoras niger				*Oxydoras kneri*
cuiu-cuiu				armado chancho
Ageneiosidae				
Ageneiosus brevifilis	*Ageneiosus* sp.	*Ageneiosus ogilviei*		*Ageneiosus brevifilis*
mandube	bagre zapato	mandube		manduba
Ageneiosus ucayalensis				*Ageneiosus valenciennesi*
patinha				manduvi
Hypophthalmidae				
Hypophthalmus edentatus	*H. edentatus*	*H. edentatus*		
mapará	bagre paisano	highwaterman		
Hypophthalmus perporosus				
mapara				
Loricariidae				
Hypostomus [*Plecostomus*] spp.		*Hypostomus* spp.	*Hypostomus regani*	
acari, bodo		suwerick	cascudo	
Pterygoplichthys pardalis			*Hypostomus albopunctatus*	
acari			cascudo	
Callichthyidae				
Hoplosternum littorale	*Callichthys callichthys*	*Hoplosternum littorale*		
tamoatá	curito, busco	hasser		
Atherinidae				*Basilichthys bonariensis*
				pejerrey
Sciaenidae				
Plagioscion squamosissimus	*P. squamosissimus*	*Plagioscion* sp.		
pescada	curvinata	basha		
Plagioscion sp.	*Plagioscion* sp.			
pescada vermelho	curvinata negra			
Cichlidae				
Cichla ocellaris	*Cichla ocellaris*	*Cichla ocellaris*		
tucunare	pavon	lukanani		
Cichla temensis	*Cichla temensis*			
tucunare paca	pavon			
Astronotus ocellatus	*Astronotus ocellatus*			
acará açu	pavona, cupaneca			
Cichlasoma spp.	*Cichlasoma* spp.	*Cichlasoma* spp.		
acará	mochoroca	patwa		
Crenicichla spp.	*Crenicichla* spp.	*Crenicichla* spp.		
jacunda	mataguaro	jacarunde		

and doradids as they moved upriver, and (3) from gillnetting fishes feeding in the flooded forest of the Rio Machado and other tributaries of the Madeira. In Porto Velho the large pimelodid catfishes made up 21% of the fish passing through the fish market. Many of these, however, were exported from the region directly to Southeast Brazil. Of the characoids, prochilodontids (23% of the weight sold) were here the most important, followed by *Brycon* (16%); *Colossoma* made up only 0.6% of the catch. *Prochilodus nigricans* is reported by the fishermen to be the most abundant food fish in the Rio Guaporé and Rio Mamoré, which flow into the Rio Madeira (Goulding 1981).

In the Itacoatiara area, 250 km down the Amazon river from Manaus, fish were caught by about twelve different fishing methods. Smith (1981), who analyzed the market data, also recorded the habitats of the 86 different species caught, whether from the main river, side channels, lakes or flooded forest. The prochilodontids were by far the most important group, 67% of the total catch of 4701 kg. *Osteoglossum* and the loricariids were next in importance, while *Colossoma* constituted only 1.9% of the weight landed here.

Experimental gillnets set at ten Central Amazon sites (four lakes, three of them near Manaus plus Lake Amana upriver on the tributary Japurá, and six sites on the Blanco and Negro north of Manaus) produced over 80 species of fish. The main biomass of these was of piscivores (Barthem 1981). Comparison of catches in the three floodplain lakes near Manaus suggested that the communities are strikingly similar in terms of species but that there are great differences in the relative abundances from lake to lake. Gillnet mesh size selectivity curves were determined for *Pellona*, *Plagioscion*, *Hemiodopsis* and *Curimata* species. The bulk of the catch was of crepuscular species, active at dawn and dusk. Nocturnal fishes, caught mainly at night, included the characoid *Rhaphiodon vulpinus*, the clupeid *Pellona flavipinnis* and the sciaenid *Plagioscion* species, while the cichlids (*Cichla ocellaris*, *Geophagus* and *Cichlasoma* species) were entirely diurnal.

On the Upper (Peruvian) Amazon and its tributaries, Chapman (1978) reported that 25 to 30 species a day appeared in Iquitos and Pucallpa fish markets, though only about nine species account for most of the catch. Catches of ~ 15 t day^{-1} were taken with peak catches at low water. In Iquitos the main commercial catches were *Prochilodus* (boquichicos, 10 to 40% of the catch), *Arapaima* (here called paiche, 10 to 30%), *Colossoma* (gamitanas, 5 to 10%), *Mylossoma* (palometa, 6 to 8%), large pimelodids (dorado, 5 to 10%) and loricariids (cachamas, 5 to 6%). At Pucallpa *Prochilodus* were again the main species (10 to 35%), followed by *Mylossoma* (7 to 12%), large pimelodids (5 to 10%), and loricariids (5 to 6%). *Arapaima* and *Colossoma* had become of little significance (reputedly overexploited). At Pucallpa, *Triportheus* species and curimatids were also important in the fish market. Other commercial species in the Upper Amazon included *Osteoglossum*, *Rhaphiodon*, *Hoplias malabaricus*, *Cichla ocellaris*, *Ancistrus* species and small tetragonopterins which exist in such numbers that there was a plan to make these into fish meal at Iquitos (see also Hanek et al. 1982b).

Orinoco. — In the Orinoco the main commercial species between Caicara and the delta were those listed in Table 1 (Novoa & Ramos 1978), namely the large characoids *Colossoma macropomum* and *C. brachypomum* (this latter especially in the delta), with *Mylossoma* of several species, migratory *Prochilodus mariae* and *Semaprochilodus laticeps*. The pimelodid catfishes included many of the same species caught in the Amazon such as *Brachyplatystoma* species, *Pseudoplatystoma fasciatum*, *Phractocephalus hemiliopterus* and smaller numbers of *Sorubimichthys planiceps* and *Pinirampus pirinampu*. Two *Plagioscion* species and another 30 or so less common species were also taken. *Pseudopimelodus* and *Callichthys* were also common in channels of the delta region.

Guyana. — In Guyana, coastal fish markets sold relatively few freshwater species, but these included *Brachyplatystoma vaillanti*, *B. filamentosum*, and *Hypophthalmus edentatus*, all caught in the estuaries in the rains; *Colossoma brachypomum* (morocot) was caught on tidal mudflats in the rainy seasons when the sea was diluted by affluent rivers (Lowe-

McConnell 1962, later identifications by Mees 1974). When seafish were scarce, the erythrinids *Hoplias malabaricus* and *Hoplerythrinus unitaeniatus*, the callichthyid *Hoplosternum littorale* and various cichlids often appeared in the markets. *Cichla ocellaris* was a valued food and sport fish throughout the country (Lowe-McConnell 1964, 1969). The large red pacu *Myleus pacu*, growing to 10 kg, is hooked in coastal rivers.

Magdalena. — In the Magdalena river of Colombia, well over half the commercial catch is said to consist of *Prochilodus reticulatus*, though *Brycon moorei*, *Pseudoplatystoma fasciatum*, *Pimelodus clarias*, *P. grosskopfii*, *Sorubim lima*, *Ageneiosus caucanus* and *Plagioscion surinamensis* also contribute significantly (Granados 1975 in Welcomme 1979).

Paraná. — In the Paraná-Paraguai system, records of catches in the Mogi Guaçu (a tributary of the Upper Paraná) have been kept for nearly forty years. Of the 95 species recorded, the main commercial fishes were the characoids *Prochilodus scrofa* and *Salminus maxillosus* with other species as listed in Table 1. In the Middle Paraná, *Prochilodus platensis* and *Salminus maxillosus* are very important species, together with *Pimelodus clarias* and *P. albicans*, *Hoplias malabaricus*, *Schizodon fasciatum* and *Leporinus obtusidens* (Bonetto et al. 1969, Bonetto 1975). In the Rosario fishery zone of the Paraná river, Vidal (1969; quoted by Welcomme 1979) lists the species of major commercial value as mainly characoids and catfishes together with the engraulid *Lycengraulis olidus* and the atherinid *Basilichthys bonariensis*.

Lake Titicaca. — The largest natural lake in South America, Lake Titicaca (7992 km^2), is high (3800 m) in the Andes. Fisheries were formerly based on the indigenous *Orestias* cyprinodontoids (eighteen species, Eigenmann & Allen 1942), but their populations declined after the introduction of rainbow trout, *Salmo gairdneri*. The main catch is now $\sim$400 t yr^{-1} of these trout (Zeisler & Ardizzone 1979, Hanek et al. 1982a).

Summary of catch statistics

The above data show the main food fishes over much of the continent to be:

1. The deep-bodied serrasalmine characoids.
2. The migratory prochilodontid species.
3. Other large characoids, particularly *Brycon*, *Salminus*, cynodontid and erythrinid species (the latter in savanna floodplain areas) and the omnivorous anostomids and hemiodids.
4. Smaller but very numerous characoids, *Triportheus* and curimatids, used mainly when larger fishes are not available.
5. Catfishes, including the large pimelodids, *Hypophthalmus* where these occur, small loricariids and callichthyids.
6. Cichlids, especially *Cichla*, very important as subsistence (and sport) fish over much of the continent.
7. The osteoglossids *Arapaima* and *Osteoglossum*.
8. The sciaenids *Plagioscion* and the clupeids *Pellona*.

Numerous other kinds (e.g. acestrorhynchine characoids, auchenipterid catfishes) are also eaten by subsistence fishermen.

Quantitative estimates of fish production in tropical freshwaters

It is much more difficult to obtain comparable quantitative data on fish populations, their biomasses, production and yields from rivers than from confined bodies of water such as lakes and reservoirs, as river fishes are so mobile. When comparing data, one has to distinguish clearly between biomasses (the standing stock of fish present in a water body at any one time), the ecological production and the catch (or yield) which represents only a proportion of the production. This proportion can range from the entire production in fish ponds, to a very small proportion where fishing conditions are difficult, where there are few fishermen or where natural mortalities are great.

Standing stock

The few published estimates of ichthyomasses from South American waters are compared in Table 2 with comparable data from other tropical freshwaters. Ichthyomass figures are mainly from floodplain lagoons or pools, fished at low water. The difficulty with comparing such estimates is that fishes from a very wide area are concentrated in pools in the dry season. Values are not, therefore, equivalent to crops from fish ponds.

There are considerable variations between different biotopes within one area. For example, on the Magdalena floodplains Kapetsky (1977) estimated biomasses between 0.23 and 251 kg ha^{-1} in the open water and between 20 and 2323 kg ha^{-1} in the bays. There was also a negative correlation between fish stock density and water level (see Welcomme 1979). Such differences showed up even more clearly from chemofished samples on the Kafue River floodplain (in the Zambezi system, Africa), where biomasses from open water and the river channel were very different from those found in vegetated lagoons (Lagler et al. 1971). In the Kafue River channel Kapetsky (1974) obtained very different values at three sites, which he thought reflected different fishing pressures in the three areas. In the Sokoto River (Nigeria), pools with a mixture of sand and mud carried higher ichthyomasses than sand or mud-bottomed pools (the latter containing mainly small fishes) (Holden 1963). In an arm of the Chari River (West Africa), the ichthyomass was greatly reduced two months after the initial sample (at the end of the flood) and was very small in the same place the following year, indicating great variation from year to year (Loubens 1969). These examples stress the need for caution when trying to make comparisons between different locations.

In the Middle Paraná near Santa Fé, fish com-

Table 2. Comparison of ichthyomass (kg ha^{-1}) in tropical freshwaters (for further data see Welcomme 1979).

	mean	(range)	Authority
South America			
Apure floodplain, Venezuela	982		Mago-Leccia 1970b
Magdalena floodplain, Colombia	122		Kapetsky 1977
open water	55.7	(0.23–251)	
bays	78.9	(20–2323)	
Paraná floodplain, Argentina			
8 temporary lagoons	1264	(175–6500)	Bonetto et al. 1969
4 permanent lagoons	918	(550–1287)	
Mogi Guaçu, Brazil	313		Gomez & Monteiro 1955
Africa			
Sokoto R. pools, Nigeria			Holden 1963
sand bottom	785	(691–1007)	
mud bottom	233	(196–270)	
intermediate	1012	(585–1440)	
Chari R., W. Africa			Loubens 1969
river arm	5116		
river 2 months later	1600		
river following year	369		
Niger R., Kainji, Nigeria	60+		Motwani & Kanwai 1970
Kafue R. channel (3 sites)	106 ± 29.2	576 ± 129.2	Kapetsky 1974
Kafue floodplain, Zambia	High water	Low water	Lagler et al. 1971
open-water lagoon	337	426	
vegetated lagoon	2682	592	
river channel	337	204	
grass marsh	64	dry	

munities in eight temporary oxbow lakes studied at low water by Bonetto et al. (1969) and Bonetto (1975) produced 41 species, representing 13 families of fishes. Characoids predominated in all of these lakes, and the number of individuals was very high. Numbers and biomasses varied very much from lake to lake; the mean biomass was over 1200 kg ha^{-1} of the dry-season area of the lake. A large proportion of the biomass was made up of relatively few species; over 30% of the total biomass was of the mud-feeding *Prochilodus platensis*. The piscivorous *Hoplias malabaricus* made up 25% of the biomass in some lakes, and the innumerable small characids (such as *Astyanax*) and curimatids over 20% of the total biomass. Other species present in numbers included cichlids (mainly *Aequidens*), abundant in oxbows with abundant plant cover, *Pimelodus clarias*, *Schizodon fasciatum*, *Leporinus obtusidens*, *Synbranchus marmoratus*, hypostomine and loricariid catfish, *Serrasalmus* and other small carnivores. Most fish were the juveniles of lotic species; juvenile *Prochilodus* dominated the biomass in many pools.

Chemofishing and capture-mark-recapture methods, used to census fish stocks in four permanent lakes in the Paraná floodplain near Santa Fé (Bonetto et al. 1969, Bonetto 1975), showed *Prochilodus platensis* to be dominant (61% of the biomass), followed by loricariids (6%), *Pimelodus clarias* (5%) and *Hoplias malabaricus* (4%). The shallowest (semipermanent) lake had the most *Hoplias*. The young of some large species such as *Salminus maxillosus* and *Pseudoplatystoma coruscans* were found in permanent lakes but not in temporary ones. The biomass was again high, over 900 kg ha^{-1} from the low-water area of these permanent lakes. The associations of species in other Paraná floodplain pools have been studied by Cordiviola de Yuan & Pignalberi (1981).

The returns of tagged fishes were used by Godoy (1975) to estimate the stocks of fishes in the Mogi Guaçu system (using Petersen's formula $P = m(u + r)/r$, when P = stock, m = number of marked fish, u = number of unmarked fish from catch statistics, r = number of recaptures). This fish stock was estimated to be 3,440,000 ($\sim$3440 t of fish between 25.0 and 80.0 cm TL). This was taken to represent $\sim$800 kg km^{-1} of the 400 km long Mogi Guaçu River, equivalent to $\sim$133 kg hectare fluvial^{-1}, i.e. considerably less than the 313 kg ha^{-1} from the reservoirs of the Pirassungungu experimental station determined by Gomez & Monteiro (1955). (But as the total area of the three river systems which form 'one ecosystem' for these migratory fishes is given by Godoy as 39,611 km^2, this would seem to represent much less – only $\sim$1 kg ha^{-1} fish.) In the Mogi Guaçu where catch records have been kept since 1942, the *Prochilodus scrofa* catches, which form 50 to 60% of the 320 t yr^{-1}, landed near Cachoeira de Emas, have continued without any signs of diminishing. But *Brycon lundii*, a popular food fish in the early years (1606 kg landed in 1942), virtually disappeared from the catches. Godoy (1975) attributed this decline to environmental changes, particularly clearing the riverside forest which had provided food for these fish.

Standing stocks also vary greatly from year to year, as shown in Africa in the Kafue floodplain studies. Initial studies indicated that recruitment, growth and survival were better in years when river discharges were high (Kapetsky 1974). Later investigations, however, showed that the picture is more complex, tilapia year classes being higher following low-flood years (Dudley 1974, 1979). For Kafue tilapia, scale rings enabled growth to be backcalculated and variations in growth from year to year of the different year classes related to temperatures and flood levels (over 15 years). For other species severity of the dry season appeared to affect subsequent stocks more than did the intensity of flooding.

A unique opportunity to census a stretch of the Niger river was presented at Kainji (Nigeria) when half the river (here divided by an island) was enclosed between two coffer dams in May 1966, and the water was pumped out in July. The enclosed 1.7 km long stretch of river (18 ha area) produced over eighty species of fish totalling 1069 + kg (equivalent to 60 kg ha^{-1}, Motwani & Kanwai 1970). This stretch of river, with rocky walls, coarse sand bottom devoid of vegetation, appeared an inhospitable habitat for the fish and may have been used as a low-water refuge. Fifty of the trapped species had mature or ripe gonads, and many spawned. These may have been moving up or down river when trapped between the coffer dams.

Ecological production

Ecologists are concerned with 'total production', the amount of fish tissue produced per unit time whether or not it survives to the end of that time, whereas fishery biologists often concentrate on the fish flesh available for capture at the end of that time (available production and yield, Balon 1974). Very few estimates have been made for production in tropical freshwaters (none of them in South America), and these are shown in Table 4 for some Cuban lakes (Holčik 1970), for Lake Kariba (where first estimates put forward by Balon 1974, had to be revised, Mahon & Balon 1977) and for three species of tilapia on the Kafue eastern floodplain (Kapetsky 1974). The stock turnover, indicated by the production/biomass ratio, ranged from 0.7 in 20 preferred (large) species from Lake Kariba (with a much faster P/B ratio, 2.3, in smaller accompanying species in this lake) to 1.59 in *Tilapia rendalli.*

Pond culture has shown that fish growth and production vary greatly with available food and conditions affecting the efficiency of its use. In Zaire, Central Africa, 'managed' ponds, in which the water was fertilized and fish fed, produced 1000 to 4000 kg ha^{-1} yr^{-1} of tilapia, compared with but 100 to 1000 kg ha^{-1} yr^{-1} in unmanaged ponds (Huet 1957). In Israel, unfertilized carp ponds produced 90 to 100 kg ha^{-1} yr^{-1}, fertilized ponds 300 to 400 kg ha^{-1} yr^{-1} and with additional food 900 to 1500 kg ha^{-1} yr^{-1}, while a polyculture of species fed in fertilized ponds gave 2500 kg ha^{-1} yr^{-1} (Hepher 1967). In South America pond experiments have been mainly with characoid fishes; these will not spawn in still water, and pituitary injections have to be used to induce spawning to obtain juveniles for stocking (Fontenele et al. 1946). *Colossoma macropomum*, a fast-growing fish that can withstand handling, is considered a promising candidate for pond culture both in Brazil and Venezuela (Lovshin et al. 1974, Novoa & Ramos 1978). In ponds males mature at two years, females at three when 2.5 to 3.0 kg weight. Monocultures of this species in Brazil produced an average of 6683 kg ha^{-1} of fish averaging 1.5 kg with a food conversion ratio of 2.8. The addition of all male tilapia hybrids to these ponds increased fish production by a total of 2256 kg ha^{-1} compared with the monoculture of *Colossoma*, without increasing the quantity of feed or worsening the food conversion ratio (Lovshin 1982).

Catches

As shown in Table 3, fish catches from thirty-three lakes constructed as antidrought measures in northeast Brazil ranged from 68 to 136 kg ha^{-1} yr^{-1} (range of yearly mean values for different lakes, types of fish unspecified). The most productive lake gave a mean catch of 346 kg ha^{-1} yr^{-1}, with a maximum of 976 kg $ha^{-1}yr^{-1}$ in 1976. Eleven of the lakes gave a mean catch over 100 kg $ha^{-1}yr^{-1}$ (Paiva & Gesteira 1977).

The Boa Esperança reservoir which formed behind the hydroelectric barrier on the Parnaíba River (mid-north Brazil) and closed in 1969, is a 200 km long, ~43,000 ha man-made lake. This reservoir produced an estimated catch of 52 kg $ha^{-1}yr^{-1}$ (based on catches of 195,747 kg in 1971; 281,834 in 1972; 189,693 in 1973; Paiva 1976). The fisheries potential of the Parnaíba river, its tributaries, marginal lagoons, dams and estuarine zone were assessed by Paiva (1973). The Parnaíba has 80 to 100 species of fish: 34% of the reservoir catch was of branquinha (*Curimata*), 29% curimata (*Prochilodus lacustris*), 15% curvina (*Plagioscion squamosissimus*), 12% piranha (*Serrasalmus nattereri*), 6% sorubim (*Pseudoplatystoma fasciatum*), 3% arenque (*Ilisha castelnaeana*), 3% fidalgo (*Ageneiosus valenciennesi*), 2% mandube (*Hemisorubim platyrhynchos*), with smaller numbers of piaus (*Leporinus/Schizodon*), piratinga (*Brachyplatystoma filamentosum* and *B. vaillanti*), traira (*Hoplias malabaricus*) and mandis (pimelodids).

The differences in catches from numerous African lakes and reservoirs correlate well with the morphoedaphic index (MEI), bearing out the well

Table 3. Fish catches (yields kg ha^{-1} yr^{-1}) from various tropical freshwaters (for further data see Welcomme 1979).

South America		
Brazil, Amazon floodplains		
Itacoatiara region (0.9 t km^{-2})	9	Smith 1981
Madeira tributaries	52	Goulding 1981
Rio Negro	0.5	Petrere 1982
Parnaíba system, Boa Esperanca lake	52.7	Paiva 1976
Northeast: 33 man-made lakes (range of means)	68–136	Paiva & Gesteira 1977
Colombia, Magdalena floodplain	32.5	in Welcomme 1979
L. Titicaca, rainbow trout	0.5	Zeisler & Ardizzone 1979
Africa		
Zambezi R., Barotse floodplain	6.8	in Welcomme 1979
Kafue floodplain	15–19	in Welcomme 1979
Malawi, Shire R. floodplain	118–143	in Welcomme 1979
Reservoirs (7)	7.6–127.2	in Welcomme 1979
Lakes (40+)	~1.0–~500	Henderson & Welcomme 1974
Asia		
Lower Mekong R.	40.7	in Welcomme 1979
Ganges-Brahmaputra	78.2	in Welcomme 1979
Reservoirs: India (8)	2.1–187.7	in Welcomme 1979
Thailand (8)	16.8–135.6	in Welcomme 1979
Indonesia (8)	21.7–356.6	in Welcomme 1979
Floodplains, worldwide (at maximum flood)	40–60	Welcomme 1979

known qualitative impression that shallow, nutrient-rich lakes are more productive than lakes that are deep or low in nutrients (Henderson & Welcomme 1974, but see also Oglesby 1982, Ryder 1982). In Asia, the shallow waters with high conductivities in Indonesia are very productive. The exceptionally low catches from some Indian reservoirs were considered to be due to concentrating on a few species of cyprinid poorly adapted to lakes. Most of the reservoirs are populated by riverine fishes, some of which drop out of the fauna, while others thrive under the new conditions (as discussed by Balon 1974, Lowe-McConnell 1975).

Catches from reservoirs are generally comparable to, or slightly higher than, catches from floodplains. Data on catches from rivers have also been collated by Welcomme (1979). Catch statistics from rivers are often of low reliability, but for moderately to heavily fished African rivers, Welcomme (1979) related catches to area of river system, treating separately those with or without extensive floodplains. He concluded that the MEI accounted for much of the observed variation between rivers or streams of the same system. Rivers with a greater area of floodplain, not unexpectedly, produce more fish. The African rivers examined fell into two groups: (1) those with an extensive floodplain system, 2.5 to 3.8% of the river basin area, and (2) 'normal' rivers (floodplains less than 1.5% of the total basin). Fish catches from African rivers ranged from 6.8 kg ha^{-1} yr^{-1} for the Barotse floodplain of the Zambezi, and 15 to 19 kg ha g^{-1} yr^{-1} for the Kafue, to 118 kg $ha^{-1}yr^{-1}$ on the Shire (Malawi) floodplain, and from some Asian rivers from 34 to 78 kg $ha^{-1}yr^{-1}$. Welcomme's analyses suggested that floodplain catches should increase until there are about 10 fishermen km^{-2}, after which the total declined. In South America the floodplain river Magdalena catch is estimated at 32.5 kg $ha^{-1}yr^{-1}$. In the Amazon basin, Goulding (1981) estimated that the 150,000 ha Madeira floodplain (including lower courses of its tributaires) gave a catch of 52 kg ha^{-1} yr^{-1}, while Smith (1981) estimated a catch equivalent to 9 kg ha^{-1} yr^{-1} (excluding exported siluriforms) for the 30 km wide Amazon floodplain within 60 km of Itacoatiara; in the blackwater Rio Negro the catch was only 0.5 kg ha^{-1} yr^{-1} (Petrere 1982). Welcomme (1979) regards 40 to 60 kg ha^{-1} at maximum flood to be 'normal' for floodplains throughout the world.

Evidence for stock depletion in South American waters

It has been suggested, based on little data, that the following species are much less common than formerly in areas where they have been heavily fished: *Arapaima gigas* in the Central Amazon and the Rio Madeira (Goulding 1980), also of the Peruvian Amazon (Chapman 1978) and in the Rupununi District of Guyana; *Colossoma macropomum* and *C. bidens* in the Rio Madeira (Goulding 1981); *Cichla ocellaris* near Manaus (Petrere, personal communication). The decline of these catches has been attributed to overexploitation near centers of dense (and increasing) human population. Alternatively, in the Mogi Guaçu, *Brycon lundii*, once a popular food fish, has virtually disappeared, which Godoy (1975) attributed to the clearance of riverside forest which provided so much of their food; pollution of the marginal lagoon habitats of juvenile fishes has probably contributed to declines in catches in this river.

To assess what is happening to a fishery, reliable catch statistics are needed over a series of years, based on a catch-per-unit effort. For the Central Amazon fishery, where so many different types of gear are used, Petrere (1978a) examined the various units and concluded that catch per number of fishermen times number of fishing days is the most satisfactory estimator; he later concluded that the number of fishermen explained over 95% of the variance of the catches (Petrere 1982). Petrere's (1983) catch per recruit analysis of *Colossoma macropomum* in the Central Amazon implies that the tambaqui stock over the area is still underexploited, but his regional observations (Petrere 1982) indicate that exploitation is much heavier in the várzea lakes near Manaus than in remote areas. Factors such as increasing cost of motorboat fuels increase pressures on stocks near large markets.

Recorded catches from the Central Amazon basin have been compared with possible theoretical yields based on data from African rivers by Bayley (1981). Actual catches in 1977 were 85,300 t, compared with theoretical yields (assuming one can compare African and South American rivers) of 125,000 t with 'normal' floodplains, 217,000 t including 'extensive' floodplains. Bayley (1981) discusses management strategies to crop the possible additional yield, concluding that the 'optimum' strategy depends on whether one wants to provide a smaller catch of high-priced species from a less exploited system or a larger quantity of cheaper fish from an ecosystem of greatly altered species proportions but higher productivity.

Relevant studies in other tropical areas

It is interesting to compare the status of ecological studies on the freshwater food fishes in South America with what has been achieved in Africa. Studies in Africa have been easier to make: many of the food fishes live in the large lakes where they can be studied throughout their life histories much more easily than can far-ranging riverine fishes. The 'type' reference material of African species is concentrated in a few major museums – not as widely scattered as that for South American fishes. Faunal lists and identification keys are now available for most parts of Africa, providing the necessary basis for ecological work. Furthermore, in Africa there appears to have been a closer liaison between scientists studying fishes in research institutions and biologists concerned with fishery development in the field. In Africa, too, opportunities have been taken during the creations of the large new man-made lakes behind hydroelectric dams (such as lakes Kariba, Volta, Kainji and Nasser) to study the life histories of the fishes and how riverine fish communities are changed to lacustrine ones. A great deal of basic information has resulted from the International Biological Programme studies on productivity in freshwaters such as Lake Chad (see Lévêque 1979, Lauzanne 1976, also Albaret 1979,

Mérona & Ecoutin 1979), Lake George in Uganda (see Burgis et al. 1973), Lake Chilwa (Malawi, see Kalk et al. 1979) and Lake Sibaya in South Africa (see Bruton 1979). Floodplain studies in Africa have shown how much growth rates and breeding success vary from year to year with environmental conditions, particularly water level (see Dudley 1979).

There are now good autecological studies of commercial species, for example, on the population dynamics of the characid *Alestes baremose* (Durand 1978). On the other hand, there has been far less study of fish movements in African rivers, no tagging experiments comparable with those on the Paraná (but see Daget 1957, Matthes 1964), and most of the new dams in Africa lack fish passes.

'Warnings' from African studies include the collapse of the fishery for *Labeo altivelis*, a cyprinid once caught in great numbers on its spawning migration up the Luapula river from Lake Mweru (Kimpe 1964). *Labeo victorianus* was also overexploited as it migrated up rivers from Lake Victoria (see Fryer 1973). The unexpected effects of introduced tilapia species on the endemic tilapias of Lake Victoria (described by Welcomme 1966) stress how difficult it is to foresee the effects of introducing exotic species.

For lakes Chad and George, attempts were made to measure transfers of energy from one trophic level to the next. This proved very difficult to do as so many fishes take food from several trophic levels, or change as they grow, or seasonally. For example, 100 calories of phytoplankton in Lake Chad would allow 5.2 cals of energy to be accumulated by the top predator *Lates niloticus* by one foodpath but only 1.7 cals by another (see Lévêque 1979). Studies in the shallow equatorial Lake George, with its rich sustained fishery of phytoplankton-feeding tilapia (*Oreochromis niloticus*) suggested that despite the very high gross primary production in this lake, surprisingly little, a mere 0.8% of the estimated net primary production, gets transferred to these fishes (Burgis 1978, Burgis & Dunn 1978). The reasons for this apparent inefficiency are not clear. Zooplankton was found to be more important than fish in the turnover of nutrients in this lake.

The Lake Chilwa studies were concerned with the decline and recovery of fish stocks as this shallow lake dried up and then refilled, a phenomenon later studied in Lake Chad and other places in the Sahelian zone (see Daget 1957, 1975). In Lake Sibaya, the bottom deposits on which the tilapia (*Oreochromis mossambicus*) feed were found to be less nutritious in the deeper water into which the fish move as they grow (Bowen 1979a); Bowen has now looked at this phenomenon in Lake Valencia

Table 4. Ecological production, yields and turnover rates (P/B ratios) in some tropical freshwaters.

	Mean biomass kg ha^{-1} B	Production kg ha^{-1} yr^{-1} Total A	Available P	Yield kg ha^{-1} yr^{-1} $Y_{(A)}$	$Y_{(P)}$	P/B ratio
Cuba (Holčik 1970)						
L. Sabanilla	321		220			1.46
L. Luisa	325		276			1.18
Africa						
Kariba (Mahon & Balon 1977) (20 species)	827	1224	720	400	202	0.7
Kafue eastern floodplain (Kapetsky 1974)						
Tilapia rendalli	125	198	110	18	8	1.59
Oreochromis macrochir	145	145	96	39	23	1.0
Oreochromis andersoni	147	119	92	23	15	0.75

in Venezuela (Bowen 1979b, 1981), and his findings are very relevant for growth studies of other detritivorous fishes.

Conclusions

In South America, unlike Africa, the commercial freshwater fisheries depend mainly on migratory fishes. Tagging experiments have stressed how mobile these fishes are. There is as yet very little quantitative information available on growth rates, biomasses, catches and stock turnover rates, needed to assess fish production. The information suggests that stocks of the large species have become, or are becoming, depleted near the larger centers of human populations, but catch statistics have not been kept for long enough to prove the changes. Clearing the forest is said to have caused the decline of *Brycon lundii* in the Mogi Guaçu; forest clearance is a threat to the numerous species now known to rely on the flooded forest for foods (Goulding 1980).

In Africa stocks of riverine fishes have been found to recover fairly rapidly after drastic droughts, which suggests that South American stocks should be able to recover from overexploitation, where this has not continued for too long, if pressure is relieved. However, environmental changes such as forest clearance and pollution of the marginal lagoon habitats of the juvenile fishes, appear to present far greater hazards. The complexity of the South American fish fauna and the multispecific nature of the fisheries call for different management techniques from those developed for single species stocks in temperate regions (as discussed by Gulland 1975, Pauly 1980), but the prime need is for quantitative information on what is happening to the fish stocks, and for this reliable catch statistics over a long series of years are essential.

Acknowledgements

I am very grateful to the Trustees of the British Museum (Natural History) for working space and excellent library facilities, and to members of the Fish Section there who have helped these studies in so many ways. My special thanks go to Tom Zaret, who suggested this paper and for his great help in editing it, and to the other reviewers, Mary Power, Michael Goulding and Miguel Petrere, Jr., who have made so many helpful suggestions and given so freely of their own experience with these fishes.

References cited

Albaret, J.J. 1979. Revue des recherches enterprises sur la fécondité des poissons d'eau douce africains. Réunion Travail Limnologie Africaine, Nairobi, December 1979, ORSTOM, Paris 1–67.

Balon, E.K. 1974. Fish production of a tropical ecosystem. pp. 249–676. *In*: E.K. Balon & A.G. Coche (ed.) Lake Kariba: A Man-made Tropical Ecosystem in Central Africa, Monogr. Biolog. 24, Dr W. Junk Publishers, The Hague.

Barthem, R.B. 1981. Experimental studies on gill-net fishing in Central Amazon lakes. M.Sc. Thesis, INPA, Manaus. 84 pp. (In Portuguese).

Bayley, P.B. 1973. Studies on the migratory characin *Prochilodus platensis* Holmberg 1889. J. Fish Biol. 5: 25–40.

Bayley, P.B. 1981. Fish yields from the Amazon in Brazil: a comparison with African river yields and management possibilities. Trans. Amer. Fish. Soc. 110: 351–359.

Böhlke, J.E., S.H. Weitzman & N.A. Menezes. 1978. The status of systematic studies of South American freshwater fishes. Acta Amazonica 8: 657–677. (In Portuguese).

Bonetto, A. 1975. Hydraulic regime of the Paraná River and its influence on ecosystems. Ecol. Stud. 10: 175–197.

Bonetto, A., W. Dione & C. Pignalberi. 1969. Limnological investigations on biotic communities in the Middle Paraná river valley. Verh. int. Verein. Limnol. 17: 1035–1050.

Bowen, S.H. 1979a. A nutritional constraint in detritivory by fishes: the stunted population of *Sarotherodon mossambicus* in L. Sibaya, S. Africa. Ecol. Monogr. 49: 17–31.

Bowen, S.H. 1979b. Determinants of the chemical composition of a peripheral detrital aggregate in a tropical lake (L. Valencia, Venezuela). Arch. Hydrobiol. 87: 166–177.

Bowen, S.H. 1981. Digestion and assimilation of periphytic detrital aggregate by *Tilapia mossambica*. Trans. Amer. Fish. Soc. 110: 239–245.

Bruton, M. 1979. The fishes of Lake Sibaya. pp. 162–245. *In*: B.R. Allanson (ed.) Lake Sibaya, Monogr. Biolog. 36, Dr W. Junk Publishers, The Hague.

Burgis, M.J. 1978. Case studies of lake ecosystems at different latitudes: the tropics. The L. George lake ecosystem. Verh. int. Verein. Limnol. 20: 1139–1152.

Burgis, M.J., J.P.E. Darlington, I.G. Dunn, G.G. Ganf, J.G. Gwahaba & L.M. McGowan. 1973. The biomass and distribution of organisms in L. George, Uganda. Proc. roy. Soc. Lond. B, 184: 271–298.

Burgis, M.J. & I.G. Dunn. 1978. Production in three contrasting ecosystems. pp. 137–158. *In*: S.D. Gerking (ed.) Ecology of Freshwater Fish Production, Blackwells, Oxford.

Chapman, M.D. 1978. Ecological management strategies for Amazonian fisheries. D. Phil. Thesis, University of Oxford, Oxford.

Cordiviola, de Yuan E. & C. Pignalberi. 1981. Fish populations in the Paraná River 2. Santa Fé and Corrientes areas. Hydrobiologia 77: 261–272.

Daget, J. 1957. Données recentes sur la biologie des poissons dans le delta central du Niger. Hydrobiologia 9: 321–347.

Daget, J. 1975. Biology of the Sahelian fisheries. pp. 18–22. *In*: CIFA/OP No. 4, Annex 6, FAO, Rome.

Dudley, R.G. 1974. Growth of tilapia of the Kafue floodplain, Zambia: predicted effects of the Kafue Gorge dam. Trans. Amer. Fish. Soc. 103: 281–291.

Dudley, R.G. 1979. Changes in growth and size distribution of *Sarotherodon macrochir* and *S. andersoni* from the Kafue floodplain, Zambia, since construction of the Kafue Gorge dam. J. Fish. Biol. 14: 205–223.

Durand, J.R. 1978. Biologie et dynamique des populations d'*Alestes baremose* (Pisces, Characidae) du bassin Tchadien. Trav. Doc. ORSTOM (Paris) 98: 1–333.

Eigenmann, C.H. & W.R. Allen. 1942. Fishes of western South America. University of Kentucky Press, Lexington. 494 pp.

Fontenele, O., E.C. Camacho & R.S. Menezes. 1946. Obtaining three annual spawns of curimáta (*Prochilodes* sp., Characidae, Prochilodinae) by hormonal injections. Bolm. Mus. nac. Rio de J., Zool. 53: 1–9 (In Portuguese).

Fryer, G. 1973. The L. Victoria fisheries: some facts and fallacies. Biological Conservation 5: 304–308.

Godoy, M.P. de 1975. Characoid fishes of the Mogi Guaçu, Brazil. Editoria Franciscam, Piracicaba (4 vols.) (In Portuguese).

Gomez, A.L. & F.P. Monteiro. 1955. Population study of fishes in the reservoir of the Experimental Station of Biology and Fish Culture in Pirassungunga, Sao Paulo. Rev. Biol. Mar. Valp. 6: 82–154. (In Portuguese).

Goulding, M. 1979. Ecology of the fisheries of the Madeira River. Con. Nac. Des. Cient. Tech. (INPA), Manaus, Brazil. 172 pp. (In Portuguese).

Goulding, M. 1980. The fishes and the forest: explorations in Amazonian natural history. University of California Press, Los Angeles 280 pp.

Goulding, M. 1981. Man and fisheries on an Amazon frontier. Developments in Hydrobiology 4, Dr. W. Junk Publishers, The Hague. 137 pp.

Greenwood, P.H., D.E. Rosen, S.H. Weitzman & G.S. Myers. 1966. Phyletic studies of teleostean fishes, with a provisional classification of living forms. Bull. Amer. Mus. nat. Hist. 131: 339–455.

Gulland, J.A. 1975. Problems in multi-species fisheries. pp. 105–110. *In*: Advisory Committee on Marine Resources Research, Sesimbra, Portugal, September 1975, FAO Fish. Rept. 171 Suppl. 1.

Hanek, G. et al. 1982a. Fisheries in Lake Titicaca (Peru): present and future. FI:DP/PER/76/022, Documento de Campo 1, FAO, Rome. (In Spanish).

Hanek, G. et al. 1982b. Fisheries in the Peruvian Amazon: present and future. FI:DP/PER/76/022, Documento de Campo 2, FAO, Rome. (In Spanish).

Hepher, B. 1967. Some biological aspects of warmwater fish pond management. pp. 417–428. *In*: S.D. Gerking (ed.) The Biological Basis of Fish Production, Blackwell, Oxford.

Holden, M.J. 1963. The populations of fish in dry season pools of the R. Sokoto. Fishery Publ. Colon. Off. HMSO, London 19: 1–58.

Henderson, H.F. & R.L. Welcomme. 1974. The relationship of yield to morphoedaphic index and numbers of fishermen in African inland waters. CIFA Occas. Pap. 1: 1–19.

Holčík, J. 1970. Standing crop, abundance, production and some ecological aspects of fish populations in some inland waters of Cuba. Vestnik Čs. spol. zool. 34: 184–201.

Honda, E.M.S., C.M. Correa, F.P. Castelo & E.A. Zapelini. 1975. General aspects of fishing in the Amazon. Acta Amazonica 5: 87–94. (In Portuguese).

Huet, M. 1957. Dix années de piscicultura au Congo Belge et au Ruanda-Urundi. Compte rendu de mission piscicole. Trav. Sta. Rech. Eaux Forêsts, Groenendaal-Hoeilaat ser. D. 22: 1–109.

Kalk, M., A.J. McLachlan & C.H. Williams (ed.). 1979. Lake Chilwa: studies and change in a tropical ecosystem. Monogr. Biolog. 35, Dr W. Junk Publishers, The Hague. 462 pp.

Kapetsky, J.M. 1974. The Kafue river floodplain: an example of preimpoundment potential for fish production. pp. 497–523. *In*: E.K. Balon & A.G. Coche (ed.) Lake Kariba: A Man-Made Tropical Ecosystem in Central Africa. Monogr. Biolog. 24, Dr W. Junk Publishers, The Hague.

Kapetsky, J.M. 1977. Some ecological aspects of the shallow lakes of the Magdalena floodplain, Colombia. Paper presented to the IV International Symposium of Tropical Ecology, March 1977, Panama.

Kimpe, P. de. 1964. Contribution a l'étude hydrobiologique du Luapula-Moero. Ann. Mus. r. Afr. centr., 8vo, Sci. Zool. 128: 1–238.

Lagler, K.F., J.M. Kapetsky & D.J. Stewart. 1971. The fisheries of the Kafue flats, Zambia, in relation to the Kafue Gorge dam. Univ. Michigan Tech. Rept. FAO Rome No. FI:SF/ZAM 11 Tech. Rept. 1: 1–161.

Lauzanne, L. 1976. Regimes alimentaires et relations trophiques des poissons du lac Tchad. Cah. ORSTOM. Hydrobiol. 10: 267–310.

Lévêque, C. 1979. Biological productivity of Lake Chad. Paper for Reunion Trav. Limnologie Africaine, Nairobi, December 1979, ORSTOM Paris: 1–30.

Loubens, G. 1969. Etude de certains peuplements ichthyologiques par des pêches au poison (1re note). Cah. ORSTOM Hydrobiol. 3: 45–73.

Lovshin, L.L. 1982. Tilapia hybridization. pp. 279–308. *In*: R.V. Pullin & R.H. Lowe-McConnell (ed.) The Biology and Culture of Tilapias, ICLARM Confr. Proc. 7, Manila.

Lovshin, L.L., A.B. de Silva, J.A. Fernandes & A. Carneiro-Sobrinho. 1974. Preliminary pond culture tests of pirapitinga (*Mylossoma bidens*) and tambaqui (*Colossoma bidens*) from the Amazon river basin. FAO Aquaculture Confr., Montevideo.

Lowe-McConnell, R.H. 1962. Notes on the fishes in Georgetown fish markets and their seasonal fluctuations. Fish. Bull. 4, Dept. Agriculture, Georgetown, Guyana. 31 pp.

Lowe-McConnell, R.H. 1964. The fishes of the Rupununi savanna district of British Guiana, South America. I. Ecological groupings of fish species and effects of the seasonal cycle on the fish. J. Linn. Soc. (Zool.) 45: 103–144.

Lowe-McConnell, R.H. 1969. The cichlid fishes of Guyana, S. America, with notes on their ecology and breeding behaviour. Zool. J. Linn. Soc. 48: 255–302.

Lowe-McConnell, R.H. 1975. Fish communities in tropical freshwaters. Longman, London. 337 pp.

Lowe-McConnell, R.H. & G.J. Howes. 1981. Pisces. pp. 218–229. *In*: S.H. Hurlbert, G. Rodriguez & N.D. Santos (ed.). Aquatic Biota of Tropical South America, 2: Anarthropoda, San Diego State University, San Diego.

Mago-Leccia, F. 1970a. A list of the fishes of Venezuela, including a preliminary study of the country's ichthyofauna. Min. Agr. Cria. Of. Nac. Pesca, Caracas. 275 pp. (In Spanish).

Mago-Leccia, F. 1970b. Preliminary studies on the ecology of the fishes of Venezuela's llanos region. Acta Biol. Venez. 7: 73–102. (In Spanish).

Mago-Leccia, F. 1972. Studies on the systematics of the family Prochilodontidae with a synopsis of the fishes of Venezuela. Acta Biol. Venez. 8: 35–96. (In Spanish).

Mahon, R. & E.K. Balon. 1977. Fish production in L. Kariba, reconsidered. Env. Biol. Fish 1: 215–218.

Matthes, H. 1964. Les poissons de lac Tumba et de la région d'Ikela. Etude systématique et écologique. Ann. Mus. r. Afr. centr., Zool. 126: 1–204.

Mees, G.F. 1974. The Auchenipteridae and Pimelodidae of Suriname. Zool. Verh. (Leiden) 132: 1–256.

Merona, B. & J.M. Ecoutin. 1979. La croissance des poissons d'eau douces africains: revue bibliographique et essai de généralisation. Paper for Reunion Trav. Limnologie africaine, Nairobi, December 1979, ORSTOM, Paris: 1–139.

Miles, C. 1947. The fishes of the Rio Magdelena. Min. Econ. Nac. Soc. Piscicultura, Bogota. 214 pp. (In Spanish).

Miles, C. 1973. Economic and ecological studies on the freshwater fishes of the valle del Cauca. Cespedesia (Bol. Cient. Dept. Valle del Cauca, Colombia) 2: 9–63. (In Spanish).

Motwani, M.P. & Y. Kanwai. 1970. Fish and fisheries in the cofferdammed right channel of the R. Niger at Kainji. pp. 27–48. *In*: S.A. Visser (ed.) Kainji Lake Studies, Vol. 1, Ecology, Nigerian Inst. Soc. Econ. Research, Ibadan University Press, Ibadan.

Novoa, D.F. & F. Ramos. 1978. The commercial fisheries of the Orinoco River. Corp. Ven. Guay. Bolivar Proj. Pes. 161 pp. (In Spanish).

Oglesby, R.T. 1982. The morphoedaphic index symposium-overview and observations. Trans. Amer. Fish. Soc. 111: 171–175.

Paiva, M.P. 1973. Fish resources and fisheries in the Río Parnaíba (Brazil). Vol. Cear. Agron. (Fortaleza, Ceara) 14: 49–82. (In Portuguese).

Paiva, M.P. 1976. The fisheries of the Boa Esperança reservoir. Rev. Bras. Ener. Elet. 33: 49–56. (In Portuguese).

Paiva, M.P. & T.C.V. Gesteira. 1977. Fish production in the areas of public access in Northeast Brazil. Cen. Tech. For. Trop. 14; 55–67. (In Portuguese).

Pauly, D. 1980. A selection of simple methods for the assessment of tropical fish stocks. FAO Fish. Circ. 729. 54 pp.

Petrere, M. Jr. 1978a. Fishing and fishing pressure in the State of Amazonas. I. Per capita fishing pressure. Acta Amazonica 8: 439–454. (In Portuguese).

Petrere, M. Jr. 1978b. Fishing and fishing pressures in the State of Amazonas. II. Locations, capture methods, and landing statistics. Acta Amazonica 8: Supplemento 2: 1–54. (In Portuguese).

Petrere, M. Jr. 1982. Ecology of the fisheries in the River Amazon and its tributaries in the Amazonas State (Brazil). Ph.D. Thesis, University of East Anglia.

Petrere, M. Jr. (1983). Yield per recruit of the tambaqui (*Colossoma macropomum* Cuvier, 1818) in the Amazonas State, Brazil, J. Fish Biol. 22: 133–144.

Ringuelet, R.A., R.H. Aramburu & A.A. de Aramburu. 1967. The freshwater fishes of Argentina. Librart, SRL, Buenos Aires. 602 pp. (In Spanish).

Ryder, R.A. 1982. The morphoedaphic index-use, abuse and fundamental concepts. Trans. Amer. Fish Soc. 111: 154–164.

Saul, W.G. 1975. An ecological study of fishes at a site in Upper Amazonian Ecuador. Proc. Acad. Nat. Sci. Philad. 127: 93–134.

Smith, N.J.H. 1981. Man, fishes, and the Amazon. Columbia University Press, New York. 180 pp.

Soares, M.G.M. 1979. Ecological aspects (food and reproduction) of the port igarapé, Aripuaná, MT. Acta Amazonica 9: 325–352. (In Portuguese).

Welcomme, R.L. 1966. Recent changes in the stocks of *Tilapia* in L. Victoria. Nature, Lond. 212: 52–54.

Welcomme, R.L. 1979. Fisheries ecology of floodplain rivers. Longman, London. 317 pp.

Ziesler, R. & G.D. Ardizzone. 1979. The inland waters of Latin America. FAO, Rome, COPESCAL Tech. Pap. 1: 1–171.

Epilogue

The international spirit that was generated during the 1982 symposium on Systematics and Evolutionary Ecology of Neotropical Freshwater Fishes is clearly continuing. In October, following the meeting, I received a letter from Antonio Machado formally inviting North American scientists to participate in the II Simposio Internacional de Sistematica y Ecologia Evolutiva de Peces de Agua Dulce Neotropicales. This will take place in the IXth Latin American Congress of Zoology, being held in Arequipa, Peru, during 9–15 October, 1983. Let us hope that many North Americans will be able to attend and that these meetings will continue the spirit of scientific cooperation with our Latin American colleagues that was begun in DeKalb, Illinois, 1982.

Thomas M. Zaret
Seattle, Washington
December, 1982

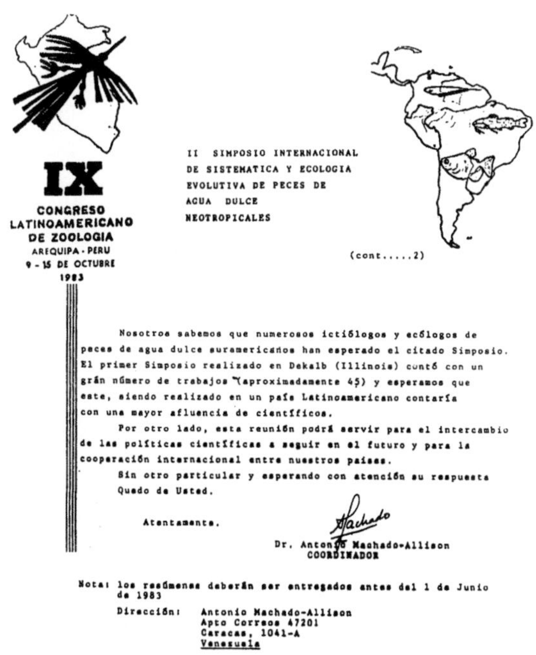
IX
CONGRESO
LATINOAMERICANO
DE ZOOLOGIA
AREQUIPA - PERU
9 - 15 DE OCTUBRE
1983

II SIMPOSIO INTERNACIONAL
DE SISTEMATICA Y ECOLOGIA
EVOLUTIVA DE PECES DE
AGUA DULCE
NEOTROPICALES

(cont.....2)

Nosotros sabemos que numerosos ictiólogos y ecólogos de peces de agua dulce suramericanos han esperado el citado Simposio. El primer Simposio realizado en Dekalb (Illinois) contó con un grán número de trabajos (aproximadamente 45) y esperamos que este, siendo realizado en un país Latinoamericano contaría con una mayor afluencia de científicos.

Por otro lado, esta reunión podrá servir para el intercambio de las políticas científicas a seguir en el futuro y para la cooperación internacional entre nuestros paises.

Sin otro particular y esperando con atención su respuesta Quedo de Usted.

Atentamente,

Dr. Antonio Machado-Allison
COORDINADOR

Nota: los resúmenes deberán ser entregados antes del 1 de Junio de 1983

Dirección: Antonio Machado-Allison
Apto Correos 47201
Caracas, 1041-A
Venezuela

Thomas M. Zaret (ed.), Evolutionary ecology of neotropical freshwater fishes. ISBN 90 6193 823 6

Ecologìa evolutiva de peces neotropicales de agua dulce

Ingestion de escamas en characoides y otros peces

Ivan Sazima, *Departamento de Zoologia, Universidade Estadual de Campinas, 13100 Campinas, Sao Paulo, Brasil*

La ingestión de escamas es conocida en varios grupos de peces no relacionados entre sí. Se conoce poco a cerca de los hábitos de la mayoría de estas especies. Los hábitos generales y el comportamiento alimenticio de algunos characoides lepidófagos son presentados y comparados con los de otras especies comedoras de escamas. La diversidad de la morfología, los hábitos y el comportamiento de los peces comedores de escamas son amplios y pocos patrones son compartidos entre los lepidófagos especializados. Excepto por la modificación de los dientes, ninguna característica morfológica permite identificar un pez como un lepidófago especializado. Las tácticas de caza consisten principalmente de acecho, emboscamiento u ocultamiento (por mimetismo agresivo). La remoción de escamas puede ser realizada por un topetazo con el hocico, generalmente dirigido a los flancos de la presa o por mordizcos o raspados. La manera de hacer la remoción de escamas parece reflejar primeramente la disposición de las mandíbulas y dientes. Las escamas son ingeridas si son tomadas directamente con la boca, si no, ellas son recogidas a medida que se hunden o tomadas desde el fondo. La lepidofagia probablemente es un hábito limitado por el tamaño. Los comedores de escamas especializados rara vez exceden los 200 mm de longitud, estando la mayoría alrededor de los 120 mm. Algunas especies comen escamas solo cuando son jóvenes; la mayoría ingiere otros alimentos además de las escamas. Los hábitos lepidofágicos probablemente se originaron a partir de comportamientos tróficos o sociales. Estos no son mutualmente exclusivos y, por lo tanto, pueden haber actuado juntos durante la evolución de la lepidofagia. Los sugeridos orígenes tróficos, incluyen el raspado de algas epilíticas, una condición piscívora modificada y la necrofagia. Los orígenes sociales comprenden un comportamiento agresivo intra e interespecífico durante la alimentación.

Pastoreo de peces tropicales de agua dulce en repuesta a diferentes escalas de variacion del alimento

Mary E. Power, *Division of Environmental Studies, University of California, Davis, CA 95616, U.S.A.*

Los peces que pastorean en riachos neotropicales se enfrentan con variaciones en su alimento – algas bentónicas arraigadas – que pueden comprender desde diferencias de calidad entre las celulas algales hasta diferencias en la produccion primaria de los habitat disponibles. Los peces pueden responder a variaciones en algunas de estas escalas pero no en otras. Por ejemplo, los bagres Loricariidae de un riacho de Panamá respondieron muy estrechamente a variaciones en la productividad de las algas en distintas pozas del riacho. En pozas soleadas, donde las algas crecían aproximadamente seis a

Thomas M. Zaret (ed.), Evolutionary ecology of neotropical freshwater fishes. ISBN 90 6193 823 6

siete veces más rápido que en pozas sombreadas, las poblaciones de loricaridos eran entre seis y siete veces más densas. En consecuencia, los indices de crecimiento de individuos pre- reproductivos de *Ancistrus spinosus* (la especie mas comun en las pozas) fueron similares en pozas con diferente dosel, correspondiendo con las predicciones de la 'hipotesis ideal de distribución libre'. Pero en pequeña escala, dentro de las pozas, le elución de aves predadoras y de predadores terrestres superó a las consideraciones de pastoreo, Las especies mas grandes, y las clases de tamaño mayores eludieron las aguas con profundidad menor de 20 cm, donde como resultado se dieron los unicos casos de algas bentonicas con biomasa suficientemente grande como para ser medida por raspado. Durante la estación seca, en que el alimento fué la limitante principal, los loricaridos presentaron mayor superposición en el uso del substrato ya que diferentes especies buscaban protección en refugios comunes, tales como troncos y ramas enredadas en las pozas. Las variaciones estacionales en el ritmo de crecimiento de los loricaridos reflejan estas restriciones.

Comunidades de peces a lo largo de gradientes ambientales en un sistema de corrientes tropicales

Paul L. Angermeier & James R. Karr, *Department of Ecology, Ethology and Evolution, University of Illinois, 606 E. Healey, Champaign, IL 61820, U.S.A.*

La estructura de la comunidad de peces en 9 corrientes boscosas (1–6 m de ancho) en la zona central de Panamá fue examinada durante las estaciones de sequía en un período de 3 años. Las regiones estudiadas variaron en la precipitación anual, el grado de sombreo por el dosel y el relieve topográfico. Invertebrados bénticos fueron mas abundantes en las corrientes que en los pozos y mas abundantes al principio de la estación seca (Enero) que a finales (Marzo). En adición, la abundancia del bentos presentó una relación negative con el sombreo producido por el dosel en las regiones estudiadas. La abundancia de invertebrados terrestres fue mayor en Enero que en Marzo y estuvo correlacionada con el ancho de la corriente. Los peces fueron asignados a 7 'guilds' alimenticios (algívoros, insectívoros acuáticos, insectívoros generales, piscívoros, comedores de escamas, herbívoros terrestres, omnívoros) en base a la similaridad de los contenidos del tracto digestivo. Cuatro especies presentaron marcados cambios de dieta con el incremento en tamaño. La distribución de los 'guilds' alimenticios (biomasa) entre los habitats y corrientes generalmente no estuvieron correlacionados con la obtenibilidad de sus mayores fuentes de alimento.

Todos los 'guilds' alimenticios, excepto insectívoros acuáticos estuvieron mas concentrados (biomasa por área) en pozos profundos. La densidad de algívoros y herbívoros terrestres aumentó con el tamaño de la corriente, pero la densidad de insectívoros acuáticos disminuyó. La riqueza en especies de los 'guilds' alimenticios aumentó con el tamaño de la corriente y la apertura del dosel. La proporción de biomasa de peces mantenidos por algas y material vegetal terrestre incrementó con el tamaño de la corriente, mientras que la mantenida por invertebrados acuáticos y terrestres declinó. Peces pequeños (<40 mm LT) fueron mas abundantes en pozos de corrientes pequeñas. La presencia de depredadores terrestres parece ser mas importante que la obtenibilidad de alimento en la distribución de peces entre los diferentes habitats. Sin embargo, la diversidad trófica de la comunidad de peces puede ser relacionada a la seguridad de conseguir una fuente de alimento disponible.

La condicion detritivora en comunidades de peces neotropicales

Stephen H. Bowen, *Department of Biological Sciences, Michigan Technological University, Houghton, MI 49931, U.S.A.*

Las comunidades de peces de gran sistema fluvial de Sur América contienen una alta proporción de peces detritívoros pertenecientes a las familias Prochilodontidae y Curimatidae. Estas familias comprenden una parte importante de los peces existentes llegando en algunos casos a constituir mas del 50% de la ictiomasa de la comunidad. Considerados como un grupo, los detritívoros tienen adaptaciones para colectar y digerir el detritus, pero los mecanismos reales de estas presuntas adaptaciones, hasta ahora, solo han sido inferidas. Los requerimientos dietéticos no han sido determinados. Adaptaciones en el comportamiento están involucradas en la selección de hábitos alimentarios, pero su importancia en el aspecto nutricional es desconocida. Es muy importante conocer y comprender la biología alimentaria de los detritívoros ya que muchos de ellos tienen importancia comercial. Además, la progresiva construcción de obras hidraúlicas amenaza interrumpir las migraciones estacionales entre las áreas de alimentación y desove.

Ecologia evolutiva de la respiracion en peces: un analisis basado en los costos de respiracion

Donald L. Kramer, *Department of Biology, McGill University, 1205 Avenue Docteur Penfield, Montreal, Quebec H3A 1B1, Canada*

Los peces usan respiración acuática unimodal, respiración aérea unimodal y un amplio rango de combinaciones bimodales de respiración acuática y aérea con la finalidad de obtener el oxígeno que ellos necesitan. Este trabajo trata de proporcionar la estructura teórica para comprender la diversidad de los diferentes tipos de respiración. Considerando el oxígeno como un recurso, se muestra que es escaso en relación con la demanda y que un suministro insuficiente limita la actividad, el crecimiento, la reproducción y finalmente la sobrevivencia de los peces. Por lo tanto, debe haber una fuerte presión selectiva para maximizar la eficiencia de la toma de oxígeno. Tanto la respiración acuática como la aérea requieren ventilación, circulación, locomoción y estructuras respiratorias. Cada uno de estos componentes requiere costos energéticos y pueden también estar sujetos a gastos en tiempo, materiales o riesgos de depredación. Si se toman en cuenta estos costos, se pone de manifiesto que la concentración de oxígeno disuelto y la distancia entre el pez y la superficie son factores ambientales determinantes. La presión de depredación y ciertas características morfológicas y de comportamiento pueden también influir. Estos factores son integrados en una teoría general de costos respiratorios, la cual permite predecir los patrones en la elección del modo de respiratión, selección de habitat, la frecuencia de los modos de respiración en diferentes habitats y las correlaciones con los habitats, las características morfológicas y de comportamiento y los modos respiratorios.

Genetica evolutiva de la diferenciacion trofica en peces goodeidos del genero *Ilyodon*

Bruce J. Turner[1], Thaddeus A. Grudzien[1], Karen P. Adkisson[2] & Matthew M. White[1], [1] *Department of Biology, Virginia Polytechnic Institute and State University, Blacksburg, VA 24061, U.S.A., and* [2] *Department of Biology, Roanoke College, Salem, VA 24153, U.S.A.*

Algunas poblaciones de una o mas especies de peces goodeidos del género *Ilyodon* en ciertos tributarios de los ríos Coahuayana y Armería (Jalisco y Colima, Méjico) son 'dicótomos' con respecto a ciertas características morfológicas que son presumiblemente adaptaciones tróficas. Una 'forma' de boca angosta (descrita como *I. furcidens* en el río Armería) es simpátrica con una 'forma' de boca ancha (denominada *I. xantusi*). Las formas son adicionalmente divergentes en el número de dientes y de branquispinas y en la coloración de los machos maduros. Otras poblaciones son esencialmente contínuas ('no dicotómicas') en estas características.

Un investigación extensivo de las alozimas de las formas simpátricas de boca ancha y angosta en cuatro localidades en el río Tule, (un tributario del río Tuxpan en la cuenca del río Coahuayana) reveló sorprendentes variaciones geográficas en la frecuencia de genes entre poblaciones, pero no diferencias entre las formas en cualquier localidad. Los datos se corresponden con la hipótesis de que las formas son componentes del mismo conjunto genético, esto es: ellos son conespecíficos. Esta hipótesis recibe adicional soporte a partir de análisis de la descencia de hembras fecundadas en el campo.

Un polimorfismo cromosomal, posiblemente envolviendo inversiones pericéntricas, existen en uno o mas sitios en el sistema del río Tule. Los individuos tienen de 0–4 cromosomas metacéntricos. La frecuencia de distribución de los fenotipos de cromosoma metacéntrico (los homologos no pueden ser distinguidos) en este sitio es divergente entre las dos formas. El polimorfismo cromosomal del *Ilyodon* del río Tule parece ser parte de un 'stepcline' en el número de cromosomas metacéntricos que están presentes en al *Ilyodon* de la cuenca del río Coahuayana.

Se mantiene como hipótesis que las formas de *Ilyodon* en el río Tule están en contacto genético, pero que la selección disruptiva actúa para eliminar individuos con fenotipos tróficos intermedios. Por un lado, al menos, la selección es suficientemente potente para mantener diferenciación cromosomal entre las formas. La base biológica de la selección natural es desconocida, pero la obtenibilidad de fuentes de alimento está implicada por evidencias circunstanciales. Las bases genéticas de la diferenciación morfológica, incluyendo la posibilidad de un componente ecofenotípico a la variación, está actualmente bajo investigación.

Seleccion natural y sexual de patrones de color en peces poecilidos

John A. Endler, *Department of Biology, University of Utah, Salt Lake City, UT 84112, U.S.A.*

En peces, los patrones de color sirven para cuatro funciones principales: (1) como defensa en contra de los depredadores; (2) para parecer menos conspicuos ante la presa; (3) como despliegue sexual ante potenciales parejas; (4) como manifestaciones de territoriedad y otras manifestaciones ante individuos de la misma especie. Estudios en guppies (familia Poeciliidae) ilustran el balance entre los efectos de la selección sexual y la elección por las hembras favoreciendo la coloración conspicua, versus la selección por depredadores favoreciendo la coloración críptica. En este trabajo se hace especial referencia al guppy, *Poecilia reticulata*, examinando el dimorfismo sexual, dicromismo sexual, patrón de color polimórfico, patrones de color relacionados con la intensidad de depredación, depredación y los efectos del background, comportamiento, selección por las hembras, carotenoides versus colores estructurales y algunos otros modelos alternativos. En adición son discutidos los patrones de coloración polimórficos en otros poecílidos (*Phalloceros caudimaculatus*, *Xiphophorus maculatus* y *X. variatus*).

Los datos muestran que diversos elementos de cualquier patrón de color dado pueden ser influenciados por diferentes modos de selección natural. En guppies la relación entre la intensividad de la depredación y el patrón de color es diferente para manchas de color estructural, de melanina y carotenoides. Los diferentes patrones de coloración tienen diversos grados de conspicuidad sobre distinto backgrounds y pueden parecer de diferente manera a depredadores y parejas de acuerdo con sus habilidades visuales. Para la familia como un todo, el dicromismo sexual está asociado con un mayor tamaño y una mayor camada, pero parece no haber relación entre el dimorfismo sexual en tamaño y el dicromismo sexual o entre el grado de dicromismo y el patrón de color polimorfo. Son discutidos los factores que influencian la evolución de los patrones de color polimorfos, los dimorfismos y dicromismos sexuales.

Gondwana y la biogeografia de los peces galaxioides neotropicales

Hugo Campos C., *Instituto de Zoologia, Universidad Austral de Chile, Casilla 567, Valdivia, Chile*

Los galaxioides (Galaxidae, Aplochitonidae, Retropinnidae y Prototroctidae), salmónidos nativos del Hemisferio Sur, tienen un patrón de distribución y origen que es controversial entre los estudiosos de la biogeografía. Hay 4 familias con 11 géneros y aproximadamente 45 especies distribuidos en Australia, Nueva Zelandia, Nueva Caledonia, Suramérica y Suráfrica. Los peces de las familias Galaxidae y Aplochitonidae tienen estrechas afinidades con la familia Salmonidae y los Retropinnidae y Prototroctidae con los Osmeridae del Hemisferio Norte. La teoría puede ser separada en eventos ecológicos (esto es: via la dispersión de la población) versus eventos históricos (esto es: vicarianza). De acuerdo a los primeros, Australia fue el centro de origen, con una dispersión posterior. Las especies se distribuyeron por invasiones sucesivas via Nueva Zelandia a través de la corriente oceánica de Australia del Este. Finalmente, como resultado del deriva de viento oeste las especies alcanzaron Suramérica y luego Suráfrica. Especial atención se ha puesto en la presencia de la fase juvenil ('whitebait') en diversas especies diadromas de *Galaxias* para explicar la dispersión marina de estas especies. En contraste la teoría histórica supone un patrón de distribución Gondwana. Esto está sustentado por estudios de la dependencia del agua dulce presente en la mayoría de los galaxioideos (por ejemplo *Galaxias maculatus*) y la amplia distribución (Suramérica, Nueva Zelandia y Australia) de los mismos. Nosotros sugerimos que la dispersión de *G. maculatus* ocurrió de acuerdo a la teoría del patrón de distribución Gondwana y no damos soporte a la teoría de dispersión. Nuestros estudios nos permiten concluir que la mejor explicación para la distribución disjunta de peces galaxoiodes fue la fragmentación que sufrió del continente de Gondwana.

De la medicion y no medicion de nicho

Thomas M. Zaret[1] & Eric P. Smith[2], [1] *Department of Zoology and Institute for Environmental Studies, University of Washington, FM-12 Seattle, WA 98195, U.S.A., and* [2] *Department of Statistics, Virginia Polytechnic Institute and State University, Blacksburg, VA 24061, U.S.A.*

Mediciones de sobreposición de nicho han sido estudíadas ampliamente por los ecólogos, pero hasta hace poco no era posible evaluar si los valores de sobreposición calculados diferían significativamente de los que podrían esperarse por simple probabilidad. Nosotros proporcionamos pruebas estadísticas y un método para calcular los intervalos de confidencia para tres medidas de sobreposición basadas en: 1) la estadística de la relación de probabilidad ('likelihood ratio statistic'); 2) la estadística de la bondad de ajuste del chi cuadrado; y 3) la estadística de Freeman-Tukey. Nosotros también proporcionamos medios de calcular intervalos de confidencia para otras 3 medidas comunmente usadas; 4) el índice del porcentaje de similaridad; 5) medida ajustada de Morisita; y 6) un índice de la teoría de información.

Comparando las mediciones pare diferentes valores de N (tamaño total de la muestra de todos los recursos) y r (número de las diferentes categorías de recursos), nosotros recomendamos un valor de N de por lo menos 100, para obtener resultados precisos. Por encima de este valor cambios de N y r afectan menos los resultados. De las 6 mediciones nosotros recomendamos la ecuación de sobreposición basada en la estadística de Freeman-Tukey, principalmente debido a que los estimados de la varianza no dependen del conjunto de datos de recursos, sino de la medida en sí. Usando nuestros métodos uno puede: (a) comparar un valor de sobreposición calculado con cualquier valor especificado (por ejemplo: difiere él de 1.00; 0.9; etc.?). (b) comparar los valores de sobreposición para las mismas especies en diferentes oportunidades (por ejemplo: día vs. noche, día uno vs. día dos, etc.).

La segunda parte de esta contribución considera el origen de la diversidad morfológica. Un modelo simple sugiere que la competencia interespecífica no es necesaria para que se produzcan diferencias específicas entre taxas estrechamente relacionadas. La diversidad de especies puede ocurrir simplemente por: 1) separación de una población en unidades genéticamente aisladas; 2) suficiente tiempo para la evolución de diferencias específicas de las especies; y 3) una condición simpátrica de la población original. Nosotros discutimos la necesidad de distinguir entre aislamientos de las poblaciones a largo plazo que conduzcan a la especiación versus adaptaciones a corto plazo de taxas estrechamente relacionadas. Solamente por evaluación del papel jugado por el tiempo nosotros podemos interpretar el significado de la diversidad morfológica.

Estado de los estudios de peces Suramericanos de agua dulce usados como alimento

Rosemary H. Lowe-McConnell, *Streatwick, Streat via Hassocks, Sussex, England*

Este papel examina los principales peces de agua dulce de Suramérica usados como alimento, basado en su biología y en las estadísticas de mercadeo. En Suramérica las pesquerías de agua dulce están basadas en peces de ríos, principalmente carácidos y siluroideos, osteoglósidos, sciánidos y clupeidos con importancia local. Unas setenta especies o grupo de especies son comercializadas; muchas de las especies grandes tienen una amplia distribución, desde el río Orinoco hasta el río Paraná.

Datos cuantitativos provenientes de los limitados estudios de cuerpos de agua dulce de Africa, Asia y Suramérica son comparados en base a biomasa actual producción y producción biológica.

El descenso de la capture de especies preferidas cerca de los grandes centros de población humana sugieren una sobreexplotación local, pero las estadísticas de capture están insuficientemente documentadas como para probar esto. La deforestación de las márgenes de los ríos puede ser una amenaza mayor para las poblaciones de peces que la sobreexplotación misma.

Ecologia evolutiva de peixes neotropicais de agua doce

Lepidofagia em caracoides e outros peices

Ivan Sazima, *Departamento de Zoologia, Universidade Estadual de Campinas, 13100 Campinas, São Paulo, Brasil*

A lepidofagia é conhecida em diversos grupos de peixes, mas há poucas informações sobre os hábitos da maioria das espécies. Os hábitos e o comportamento alimentar de alguns Caracóides lepidófagos são aqui apresentados e comparados com os de outras espécies comedoras de escamas. A diversidade de morfologia, hábitos e comportamento, é grande nos peixes que comem escamas, havendo poucos aspectos em comum entre os lepidófagos especializados. Com exceção dos dentes, modificados, não há uma característica morfológica que permita identificar um dado peixe como lepidófago especializado. Suas táticas predatórias consistem de tocaia, aproximação dissimulada, ou disfarce (na forma de mimetismo agressivo). A remoção de escamas de presa pode ser efetuada por meio de um golpe vigoroso com of focinho, dirigido principalmente ao flanco da presa, ou mordendo, ou então raspando. O modo de remover escamas parece refletir, primariamente, a disposição das maxilas e dos dentes especializados. As escamas são ingeridas diretamente quando abocadas, ou são recolhidas enquanto afundam, ou então são recolhidas sobre o fundo. A lepidofagia é, provavelmente, um hábito limitado pelas dimensões do predador. Os lepidófagos especializados raramente excedem o comprimento de 200 mm, a maioria estando próxima de 120 mm. Algumas espécies comem escamas apenas quando na fase juvenil; a maioria dos lepidófagos ingere outros tipos de alimento, além de escamas. A lepidofagia provavelmente originou-se de comportamento trófico ou social. Os dois tipos não são mutuamente exclusivos e podem, de fato, ter agido de modo complementar no decorrer da evolução de lepidofagia. As origens tróficas, sugeridas para esta evolução, incluem pastejo de algas epilíticas, piscivoria modificada e necrofagia. As origens sociais incluem comportamento agressivo intra e interespecífico durante a alimentação.

Respostas de pastejo de peixes tropicais de água doce a diferentes escales de variação no seu alimento

Mary E. Power, *Division of Environmental Studies, University of California, Davis, CA 95616, U.S.A.*

Os peixes que pastejam em riachos neotropicais defrontam-se com variação no seu alimento – algas fixas ao substrato – a qual abrange desde diferenças em qualidade entre as células das algas, até diferenças no produção primária dos habitats disponíveis aos peixes. Os peixes podem responder a variações em algumas destas escalas, mas não em outras. Por exemplo, bagres Loricariidae, num riacho do Panamá, seguiam muito de perto a variação em produtividade das algas nos poções do riacho. Nos poções ensolarados, onde as algas cresciam cerca de sete vezes mais rapidamente que nos poções sombreados, as populações de Lori-

Thomas M. Zaret (ed.), Evolutionary ecology of neotropical freshwater fishes. ISBN 90 6193 823 6

cariídeos eram seis a sete vezes mais densas. Em consequência, os índices de crescimento de indivíduos pré-reprodutivos de *Ancistrus spinosus* (a espécie mais comum nos poções) eram semelhantes em poções sob diferentes dosséis, correspondendo às previsões derivadas da hipótese da 'distribuição livre ideal'. Porém, numa escala menor, dentre os poções, a evitação de aves predadoras e predadores terrestres sobrepujou as considerações de procura de alimento. As espécies e as classes de tamanho maiores evitavam águas mais rasas que 20 cm, onde ocorriam (como resultado) as únicas safras de algas sésseis suficientemente grandes para serem medidas por raspagem. Durante a época seca, quando o alimento era limitante ao máximo, os Loricariídeos sobrepunham-se mais, no seu uso de substrato, à medida que as diferentes espécies procuravam abrigo em refúgios comuns, como troncos caídos e emaranhados de raízes, nos poções. Variação sazonal, nos índices de crescimento dos Loricariídeos ocupantes de poções, reflete estas restrições.

Comunidades de peixes ao longo de gradientes ambientais em um sistema de riachos tropicais

Paul L. Angermeier & James R. Karr, *Department of Ecology, Ethology and Evolution, University of Illinois, 606 E. Healey, Champaign, IL 61820, U.S.A.*

A estrutura da comunidade de peixes foi examinada em 9 riachos marginados por floresta (com 1–6 m de largura) no Panamá central, durante épocas secas e aolongo de um período de 3 anos. As regiões de estudo variaram em precipitação pluvial anual, grau de sombreamento do dossel e relevo. Invertebrados bentônicos foram mais abundantes nas corredeiras que nos poções e mais abundantes no período inicial da época sêca (janeiro) que no final (março). Além disso, as abundânicias do bentos mostraram correlação negativa com o sombreamento do dossel, entre as regiões de estudo. As abundâncias de invertebrados terrestres foram maiores em janeiro que em março e mostraram correlação com a largura do riacho. Os peixes foram ordenados em 7 guildas tróficas (algívoros, insetívoros de iten aquáticos, insetívoros generalistas, piscívoros, lepidófagos, herbívoros de material terrestre, onívoros), com base na similaridade dos conteúdos do tubo digestivo. Quatro espécies exibiram, com o aumento de tamanho, mudanças marcantes de dieta. As distribuições das guildas tróficas (em termos de biomassa), entre os habitats e os riachos, em geral não mostraram correlação com a disponibilidade dos seus principais recursos alimentares.

Todas as guildas tróficas, exceto a dos insetívoros de itens aquáticos, estiveram mais concentradas (em termos de biomassa por área), nos poções fundos. As densidades dos algívoros e herbívoros de itens terrestres aumentaram à medida que aumentava o tamanho do riacho, porém a densidade dos insetívoros de itens aquáticos declinous. A riqueza de espécies das guildas tróficas aumentou à medida que aumentava o tamanho do riacho e o espaçamento do dossel. A proporção da biomassa de peixes sustentada por algas e material vegetal de origem terrestre aumentou com o tamanho do riacho, ao passo que aquela sustentada pro invertebrados aquáticos e terrestres declinou. Os peixes pequenos (<40 mm CT) foram mais abundantes nos poções de riachos pequenos. O predadores terrestres pareciam ser mais importantes que a disponibilidade de alimento, para estabelecer as distribuições dos peixes nos diversos habitats. Entretante, a diversidade trófica das comunidades de peixes pode estar relacionada à constância dos recursos alimentares disponíveis.

Detritivoria em comunidades neotropicais de peixes

Stephen H. Bowen, *Department of Biological Sciences, Michigan Technological University, Houghton, MI 49931, U.S.A.*

As comunidades de peixes dos grandes sistemas fluviais, na América do Sul, contêm uma proporção alta de peixes detritívoros das famílias Prochilodontidae e Curimatidae. Estas famílias abrangem estoques importantes de peixes, que em algumas regiões compreendem mais de 50 por cento da ictiomassa da comunidade. Como um grupo, os detritívoros possuem adaptações anatômicas e fisiológicas para apanha e digestão de detrito, mas os mecanismos efetivos destas supostas adaptações tem sido, até o momento, apenas inferidas. Necessidades dietéticas não foram identificadas. A adaptação comportamental é deduzida pela escolha do habitat de alimentação, mas o seu significado para a nutrição é desconhecido. Devido ao fato de várias destas espécies terem importância comercial e porque a construção contínua de represas ameaça romper as migrações sazonais entre as áreas de reprodução e de alimentação, um conhecimento da biologia alimenter dos detritívoros torna-se importante.

A ecologia evolutiva de modo respiratório em peixes: uma análise baseada no custo da respiração

Donald L. Kramer, *Department of Biology, McGill University, 1205 Avenue Docteur Penfield, Montreal, Quebec H3A 1B1, Canada*

Os peixes utilizam respiração aquática unimodal, respiração aérea unimodal, e uma extensa variação de combinações bimodais de respiração aquática e aérea, para obter o oxigênio que necessitam. Este ensaio visa proporcionar uma estrutura teórica, na qual esta diversidade do modo respiratório possa ser entendida. A consideração do oxigênio como um recurso mostra que ele é escasso em relação à demanda, e que um suprimento inadequado limita a atividade, o crescimento, a reprodução e, finalmente, a sobrevivência. Portanto, deve haver uma forte pressão seletiva para aumentar ao máximo a eficiência da tomada de oxigênio. Tanto a respiração aquática como a aérea necessitam de ventilação, circulação, locomoção, e estruturas respiratóridas. Cada um destes componentes apresenta um custo energético e pode também estar sujeito a custos de tempo, materiais ou envolver risco de predação. A consideração destes custos revela que a concentração de oxigênio dissolvido e a distância desde a superfície são os determinantes ambientais controladores. A pressão de predação e várias características morfológicas e comportamentais podem também ter a sua influência. Estes fatores são integrados numa teoria geral de custos respiratórios, que permite previsões de padrões na partilha respiratória, seleção de habitat, a frequência de modos respiratórios em diferentes habitats, e correlações, dentre ambientes, entre as características comportamentais e morfológicas e o modo respiratório.

Genética evolutiva da diferenciação trófica em peixes goodeídeos do gênero *Ilyodon*

Bruce J. Turner[1], Thaddeus A. Grudzien[1], Karen P. Adkisson[2] & Matthew M. White[1], [1] *Department of Biology, Virginia Polytechnic Institute and State University, Blacksburg, VA 24061, U.S.A.*, *and* [2] *Department of Biology, Roanoke College, Salem, VA 24153, U.S.A.*

Algumas populações de uma ou mais espécies do gênero de peixes Goodeídeos *Ilyodon*, em certos afluentes dos rios Coahuayana e Armería (Jalisco e Colima, México), são 'dicotômicas' em relação a caracteres morfológicos, que são presumíveis adaptações tróficas. Uma 'forma' com boca estreita (descrita como *I. furcidens* no Río Armería) é simpátrica com uma 'forma' de boca larga (denominada de *I. xantusi*). Além disso, as formas divergem no número de dentes e de rastros branquiais e na coloração dos machos adultos. Outras população são essencialmente contínuas ('não-dicotômicas') nestes caracteres.

Uma investigação ampla de alozimas das formas simpátricas com boca estreita e boca larga, de quatro localidades no Río del Tule (um afluente do Río Tuxpan, na bacia do Río Coahuayana), revelou notáveis variações geográficas nas frequências gênicas entre as populações, mas não revelou diferenças entre as formas, em nenhuma localidade. Estes dados estão de acordo com a hipótese de que as formas são componentes do mesmo conjunto génico, i.e., pertencem à mesma espécie. Esta hipótese recebe um apoio adicional a partir da análise das progênies de fêmeas fecundadas na natureza.

Um polimorfismo cromossômico, provavelmente envolvendo inversões pericêntricas, existe em um ou mais locais no sistema do Río del Tule, tendo os indivíduos de 0 a 4 cromossomos metacêntricos. A distribuição de frequências dos fenótipos com cromossomos metacêntricos (os homólogos não podem ser distinguidos) neste local é divergente entre as formas. O polimorfismo cromossômico nos *Ilyodon* do Río del Tule parece fazer parte de um 'cline em degraus' no número de metacêntrricos, entre os *Ilyodon* da bacia do Río Coahuayana.

É feita a hipótese de que as formas de *Ilyodon* no Río del Tule estão em contato genético, mas que a seleção disruptiva age eliminando os indivíduos com fenótipos tróficos intermediários. Em um local, pelo menos, a seleção é suficientemente potente para promover diferenciação cromossômica entre as formas. A base biológica da seleção natural é desconhecida, mas a disponibilidade de recursos alimentares está envolvida por evidência circunstancial. A base genética da diferenciação morfológia, incluindo a possibilidade de um componente ecofenotípico para a variação, está, no momento, sendo investigada.

Seleção natural e sexual sobre padrões de coloração em peixes poecilidos

John A. Endler, *Department of Biology, University of Utah, Salt Lake City, UT 84112, U.S.A.*

Em peixes, os padrões de cores possuem quatro funções principais: (1) como defesa contra predadores; (2) para reduzir a conspicuidade do predador para a presa; (3) como exibição sexual a parceiros potenciais, e (4) como exibições, territorial e outras, para indivíduos da mesma espécie. Estudos em barrigudinhos (família Poeciliidae) ilustram o equilíbrio entre os efeitos da seleção sexual e escolha feita pela fêmea, favorecendo coloração conspícua, versus seleção por predadores, favorecendo coloração conspícua, versus seleção por predadores, favorecendo coloração críptica. O presente estudo dá ênfase ao barrigudinho, *Poecilia reticulata*, examinando dimorfismo sexual, dicromismo sexual, polimorfismo no padrão de coloração, padrões de cores relacionados a intensidade de predação, predação e os efeitos do fundo, comportamento, escolha feita pela fêmea, cores estruturais versus cores devidas a carotenóides, e alguns modelos alternativos. São discutidos também polimorfismo do padrão de cor em outros Poeciliídeos (*Phalloceros caudimaculatus*, *Xiphophorus maculatus*, e *X. variatus*).

Os dados mostram que os diferentes elementos de um dado padrão de coloração podem ser influenciados por diferentes modos de seleção natural. Nos barrigudinhos, *P. reticulata*, a relação entre a intensidade de predação e o padrão de cor é diferente para manchas devidas a melanina, carotenóides, e cores estruturais. Os diferentes padrões de coloração possuem diferentes graus de conspicuidade em diferentes fundos, e podem se apresentar de maneira diferente para predadores e parceiros com capacidades visuais diferentes. Para a família como um todo, o dicromismo sexual está associado com maiores tamanhos e maiores progênies, mas parece não haver relação entre o dimorfismo sexual em tamanho e o dicromismo sexual, ou entre o grau de dicromismo e o polimorfismo de padrão de coloração. São discutidos fatores que influenciam a evolução de polimorfismos de padrão de coloração, de dimorfismo sexual e de dicromismo sexual.

Gondwana e a biogeografia de peixes galaxióides neotropicais

Hugo Campos C., *Instituto de Zoologia, Universidad Austral de Chile, Casilla 567, Valdivia, Chile*

Os Galaxióides (Galaxiidae, Aplochitonidae, Retropinnidae e Prototroctidae), salmonides nativos do Hemisfério Sul, possuem padrão de distribuição e origem controversos entre os estudiosos de biogeografia. As famílias estão distribuidas na Austrália, Nova Zelândia, Nova Caledônia, América do Sul e África do Sul, com 4 famílias, 11 gêneros e cerca de 45 espécies. Os peixes Galaxiídeos e Aplochitonídeos possuem muita afinidade com os Salmonidae, e os peixes Retropinnídeos e Prototroctídeos com os Osmeridae, do Hemisfério Norte. A teoria pode ser dividida em eventos ecológicos (i.e., via dispersão da população) versus históricos (i.e., vicariância). De acordo com a primeira alternative, a Austrália foi o centro de origem, com dispersão subsequente, as espécies avançando por sucessivas invasões via Nova Zelândia, através da Corrente Oceânica do Leste Australiano. Finalmente, come resultado da deriva do vento oeste, as espécies alcançaram a América do Sul e, finalmente, a África do Sul. A ênfase principal para este ponto de vista, portanto, tem sido tentar explicar a dispersão, através do mar, pela presença da fase juvenil 'whitebait' em várias espécies diádromas de *Galaxias*. Por outro lado, a teoria histórica supõe um padrão de distribuição tipo Gondwana. Isto é apoiado por estudos sobre a dependência da água doce da maioria dos Galaxióides (e.g. *Galaxias maculatus*) e uma distribuição ampla (América do Sul, Nova Zelândia e Austrália) para o padrão de distribuição dos Galaxióides. Nós sugerimos que a dispersão de *G. maculatus* ocorreu após o presente padrão de distribuição dos Galaxióides, não apoiando a teoria da dispersão. Nosses estudos nos levaram a concluir que a melhor explicação para a distribuição disjunta dos peixes Galaxióides foi a ruptura que ocorreu no continente do Gondwana.

Sobre medir e não medir nichos

Thomas M. Zaret[1] & Eric P. Smith[2], [1] *Department of Zoology and Institute for Environmental Studies, University of Washington, FM-12 Seattle, WA 98195, U.S.A., and* [2] *Department of Statistics, Virginia Polytechnic Institute and State University, Blacksburg, VA 24061, U.S.A.*

As medidas de sobreposição do nicho tem sido amplamente usadas pelos ecólogos, mas até recentemente não era possível avaliar se os valores calculados de sobreposição diferiam significativamente do esperado ao acaso. Fornecemos aqui testes estatísticos e um método para calcular os intervalos de confiança para 3 medidas de sobreposição, baseados em: (1) o teste da razão de verossimilhança; (2) o teste do X^2 para aderência e (3) o teste de Freeman-Tukey. Também fornecemos um meio de calcular os intervalos de confiança para 3 outras medidas habitualmente usadas, (4) o índice de similaridade de porcentagem; (5) o índice ajustado de Morisita; e (6) um índice da teoria da informação. Comparando as medidas para diferentes valores de N (tamanho da amostra total de todos os recursos) e r (número das diferentes categorias de recursos), recomendamos um N de pelo menos 100, para a obtenção de resultados precisos. Acima deste valor, variações em N e r têm efeito reduzido sobre os resultados. Das 6 medidas, recomendamos a equação de sobreposição baseada no teste de Feeman-Tukey, principalmente porque a estimative da variância não depende do conjunto de dados dos recursos, mas da própria medida. Usando os nossos métodos, pode-se: (a) comparar um valor calculado de sobreposição com qualquer valor especificado (e.g. se é diferente de 1.00, 0.9, etc.); (b) comparar os valores de sobreposição para a mesma espécie em períodos diferentes (e.g. dia versus noite, dia um versus dia dois, etc.).

A segunda parte desta contribuição examina a origem da diversidade morfológica. Um modelo simples sugere que a competição inter-específica não é necessária para a origem de diferenças específicas entre táxons muito próximos. A diversidade de espécies pode simplesmente ocorrer a partir de: 1) divisão de uma população em unidades geneticamente isoladas; 2) período suficiente para a evolução de diferenças específicas da espécie; e 3) simpatria da população original. Discutimos aqui a necessidade de distinguir entre o isolamento da população por um prazo extenso, conduzindo à especiação, versus adaptação a curto prazo de táxons muito próximos. Somente pela avaliaçã do papel do tempo é que podemos interpretar o significado da diversidade morfológica.

O estado atual dos estudos de peixes neotropicais de água doce usados na alimentação

Rosemary H. Lower-McConnell, *Streatwick, Streat via Hassocks, Sussex, England*

Este trabalho examina os principais peixes de água doce usados na alimentação, na América do Sul, baseado em sua biologia e estatísticas de mercado. Na América do Sul, as atividades pesqueiras, em água doce, são baseadas em peixes fluviais, predominantemente os Otophysi Caracóides e Siluróides, sendo localmente importantes os Osteoglossídeos, Ciclídeos, Cienídeos e Clupeídeos. Cerca de setenta espécies, ou grupos de espécies, são comercializados; muitas das espécies grandes estão amplamente distribuídas, desde o Rio Orinoco até o Rio Paraná. Os poucos estudos feitos nos trópicos, em ambientes de água doce da África, Ásia e América do Sul, são comparados para a obtenção de dados quantitativos sobre produto em pé, rendimento e produção biológica.

Declínios na captura das espécies preferidas, nas proximidades de grandes centros de população humana, sugerem uma super-exploração local, mas as estatísticas de capture estão insuficientemente documentadas para comprová-la. Desmatamentos nas margens dos rios podem representar uma ameaça maior, no declínio do estoque de peixes, do que a super-exploração.

Species and subject index

Contents

MIX
Papier aus verantwortungsvollen Quellen
Paper from responsible sources
FSC® C105338

If you have any concerns about our products,
you can contact us on
ProductSafety@springernature.com

In case Publisher is established outside the EU,
the EU authorized representative is:
Springer Nature Customer Service Center GmbH
Europaplatz 3, 69115 Heidelberg, Germany

Printed by Libri Plureos GmbH
in Hamburg, Germany